**Advanced Courses in Mathematics
CRM Barcelona**

Centre de Recerca Matemàtica

Managing Editor:
Carles Casacuberta

For further volumes:
http://www.springer.com/series/5038

Luis A. Caffarelli
François Golse
Yan Guo
Carlos E. Kenig
Alexis Vasseur

Nonlinear Partial Differential Equations

Editors for this volume:
Xavier Cabré (ICREA and Universitat Politècnica de Catalunya)
Juan Soler (Universidad de Granada)

 Birkhäuser

Luis A. Caffarelli
Department of Mathematics
University of Texas at Austin
1 University Station, C1200
Austin, TX 78712-0257, USA

François Golse
École Polytechnique
Centre de Mathématiques L. Schwartz
F-91128 Palaiseau Cedex, France

Yan Guo
Division of Applied Mathematics
Brown University
Providence, RI 02912, USA

Carlos E. Kenig
Department of Mathematics
University of Chicago
Chicago, IL 60637, USA

Alexis Vasseur
Department of Mathematics
University of Texas at Austin
1 University Station, C1200
Austin, TX 78712-0257, USA

ISBN 978-3-0348-0190-4   e-ISBN 978-3-0348-0191-1
DOI 10.1007/978-3-0348-0191-1
Springer Basel Dordrecht Heidelberg London New York

Library of Congress Control Number: 2011940208

Mathematics Subject Classification (2010): Primary: 35Q35, 82C40, 35Q20, 35Q55, 35L70; Secondary: 86A10, 76P05, 35P25

Printed on acid-free paper

Springer Basel AG is part of Springer Science + Business Media
(www.birkhauser-science.com)

Foreword

This book contains expository lecture notes for some of the courses and talks given at the school *Topics in PDE's and Applications 2008. A CRM & FISYMAT Joint Activity*, which took place at the FisyMat-Universidad de Granada (April 7 to 11, 2008) and at the Centre de Recerca Matemàtica (CRM) in Bellaterra, Barcelona (May 5 to 9, 2008).

The goal of the school was to present some of the main advances that were taking place in the field of nonlinear Partial Differential Equations and their applications. Oriented to Master and PhD students, recent PhD doctorates, and researchers in general, the courses encompassed a number of areas in order to open new perspectives to researchers and students.

The program in the Granada event consisted of five courses taught by Luigi Ambrosio, Luis Caffarelli, François Golse, Pierre-Louis Lions, and Horng-Tzer Yau, as well as two talks given by Yan Guo and Pierre-Emmanuel Jabin. The event at the Centre de Recerca Matemàtica consisted of five courses taught by Henri Berestycki, Haïm Brezis, Carlos Kenig, Robert V. Kohn, and Gang Tian.

The volume covers several topics of current interest in the field of nonlinear Partial Differential Equations and its applications to the physics of continuous media and of particle interactions. The lecture notes describe several powerful methods introduced in recent top research articles, and carry out an elegant description of the basis for, and most recent advances in, the quasigeostrophic equation, integral diffusions, periodic Lorentz gas, Boltzmann equation, and critical dispersive nonlinear Schrödinger and wave equations.

L. Caffarelli and A. Vasseur's lectures describe the classical De Giorgi truncation method and its recent applications to integral diffusions and the quasigeostrophic equation. The lectures by F. Golse concern the Lorentz model for the motion of electrons in a solid and, more particularly, its Boltzmann–Grad limit in the case of a periodic configuration of obstacles —like atoms in a crystal. Y. Guo's lectures concern the Boltzmann equation in bounded domains and a unified theory in the near Maxwellian regime —to establish exponential decay toward a normalized Maxwellian— for all four basic types of boundary conditions. The lectures by C. Kenig describe a recent concentration-compactness/rigidity method for critical dispersive and wave equations, in both defocusing and focusing cases. The issues studied center around global well-posedness and scattering.

We are very thankful to the CRM and FisyMat for hosting these advanced courses. In particular, we thank the CRM director Joaquim Bruna for making the CRM event possible. The CRM administrative staff and FisyMat coordinators were very helpful at all times. We acknowledge financial support from the Ministerio de Educación y Ciencia, Junta de Andalucía, and Generalitat de Catalunya, institutions that supported the two events in the school.

Finally and above all, we thank the authors for their talks, expertise, and kind collaboration.

Xavier Cabré and Juan Soler

Contents

Foreword   v

1 **The De Giorgi Method for Nonlocal Fluid Dynamics**
   Luis A. Caffarelli and Alexis Vasseur   1

   Introduction   1
   1.1 The De Giorgi theorem   2
        1.1.1 Sobolev and energy inequalities   3
        1.1.2 Proof of the De Giorgi theorem   4
   1.2 Integral diffusion and the quasi-geostrophic equation 11
        1.2.1 Quasi-geostrophic flow equation   11
        1.2.2 Riesz transforms and the dependence of v on θ 12
        1.2.3 BMO spaces   12
        1.2.4 The fractional Laplacian and harmonic extensions 13
        1.2.5 Regularity 15
        1.2.6 A geometric description of the argument 22
        1.2.7 First part 23
        1.2.8 Second part 30
        1.2.9 Oscillation lemma   34
        1.2.10 Proof of Theorem 11   34
   Bibliography   37

2 **Recent Results on the Periodic Lorentz Gas**
   François Golse   **39**

   Introduction: from particle dynamics to kinetic models 39
   2.1 The Lorentz kinetic theory for electrons 41
   2.2 The Lorentz gas in the Boltzmann–Grad limit with a Poisson
        distribution of obstacles 44
   2.3 Santaló's formula for the geometric mean free path 51
   2.4 Estimates for the distribution of free-path lengths 57
   2.5 A negative result for the Boltzmann–Grad limit of the periodic
        Lorentz gas 66

2.6　Coding particle trajectories with continued fractions . . . .　71
2.7　An ergodic theorem for collision patterns . . . .　79
2.8　Explicit computation of the transition probability $P(s, h|h')$ . . . .　84
2.9　A kinetic theory in extended phase-space for the Boltzmann–Grad
　　　limit of the periodic Lorentz gas . . . .　90
Conclusion . . . .　95
Bibliography . . . .　96

3 The Boltzmann Equation in Bounded Domains
Yan Guo　　　　　　　　　　　　　　　　　　　　　　　　　　　　**101**

3.1　Introduction . . . .　101
3.2　Domain and characteristics . . . .　103
3.3　Boundary condition and conservation laws . . . .　104
3.4　Main results . . . .　105
3.5　Velocity lemma and analyticity . . . .　107
3.6　L^2 decay theory . . . .　107
3.7　L^∞ decay theory . . . .　108
Bibliography . . . .　112

4 The Concentration-Compactness Rigidity Method for Critical Dispersive and Wave Equations
Carlos E. Kenig　　　　　　　　　　　　　　　　　　　　　　　**117**

4.1　Introduction . . . .　117
4.2　The Schrödinger equation . . . .　121
4.3　The wave equation . . . .　129
Bibliography . . . .　146

Chapter 1

The De Giorgi Method for Nonlocal Fluid Dynamics

Luis A. Caffarelli and Alexis Vasseur

Introduction

In 1957, E. De Giorgi [7] solved the 19th Hilbert problem by proving the regularity and analyticity of variational ("energy minimizing weak") solutions to nonlinear elliptic variational problems. In so doing, he developed a very geometric, basic method to deduce boundedness and regularity of solutions to a priori very discontinuous problems. The essence of his method has found applications in homogenization, phase transition, inverse problems, etc.

More recently, it has been successfully applied to several different problems in fluid dynamics: by one of the authors [12], to reproduce the partial regularity results for Navier–Stokes equations originally proven in [1]; by several authors [5], to study regularity of solutions to the Navier–Stokes equations with symmetries; and by the authors, to prove the boundedness and regularity of solutions to the quasi-geostrophic equation [2].

These notes are based on minicourses that we gave at the school *Topics in PDE's 2008* in Fisymat-Granada and CRM Barcelona, as well as at schools in Ravello and Córdoba (Argentina). The structure of the notes consists of two parts. The first one is a review of De Giorgi's proof, stressing the important aspects of his approach, and the second one is a discussion on how to adapt his method to the regularity theory for the quasi-geostrophic equation.

1.1   The De Giorgi theorem

The 19th Hilbert problem consisted in showing that local minimizers of an energy functional

$$E(w) = \int_\Omega F(\nabla w)\, dx$$

are regular if $F(p)$ is regular. By a *local minimizer* it is meant that

$$E(w) \le E(w + \varphi)$$

for any compactly supported function φ in Ω. Such a minimizer w satisfies the Euler–Lagrange equation

$$\operatorname{div}(F_j(\nabla w)) = 0.$$

Already in one dimension it is clear that F must be a convex function of p to avoid having "zig-zags" as local minimizers. By performing the derivations, the equation for w can also be written as

$$F_{ij}(\nabla w)D_{ij}(w) = 0,$$

a non-divergence elliptic equation from the convexity of F.

It was already known at the time (Calderón–Zygmund) that continuity of ∇w would imply, by a bootstrapping argument, higher regularity of w. But for a weak solution, all that was known was that ∇w belongs to L^2.

If we now formally take a directional derivative of w in the direction e, $u_e = D_e w$, we find that u_e satisfies the equation

$$D_i\left(F_{ij}(\nabla w)D_j u_e\right) = 0,$$

a uniformly elliptic equation if F is strictly convex. At this point, there are two possibilities to try to show the regularity of w: either try to link ∇w in the coefficient with $u_e = D_e w$ in some sort of system, or tackle the problem at a much more basic level. Namely, forget that

$$A_{ij}(x) = F_{ij}(\nabla w)$$

is somewhat linked to the solution. Accept that we cannot make any modulus of continuity assumption on A_{ij}, and just try to show that a weak solution of

$$D_i A_{ij}(x)D_j u = 0$$

is in fact continuous.

This would imply jumping in the invariance class of the equation. All previous theories (Schauder, Calderón–Zygmund, Cordes–Nirenberg) are based on being a small perturbation of the Laplacian: A_{ij} are continuous, or with small oscillation.

But the class of uniformly elliptic equations ($I \leq A_{ij} \leq \Lambda I$) with no regularity assumptions is a scaling invariant class in itself, and never gets close to the Laplacian.

De Giorgi then studied solutions u of

$$D_i a_{ij}(x) D_j u = 0$$

with no assumption on a_{ij}, except uniform ellipticity ($I \leq a_{ij} \leq \Lambda I$), and showed that u is C^α. Applying this theorem to $(w)_e$, he solved the Hilbert problem.

Hence, we need to prove the following:

Theorem 1. *Let u be a solution of $D_i a_{ij} D_j u = 0$ in B_1 of $\mathbb{R}^N$ with $0 < \lambda I \leq a_{ij}(x) \leq \Lambda I$ (i.e., a_{ij} is uniformly elliptic). Then $u \in C^\alpha(B_{1/2})$ with*

$$\|u\|_{C^\alpha(B_{1/2})} \leq C\|u\|_{L^2(B_1)},$$

where $\alpha = \alpha(\lambda, \Lambda, n)$.

Proof. The proof is based on the interplay between the Sobolev inequality, which says that $\|u\|_{L^{2+\varepsilon}}$ is controlled by $\|\nabla u\|_{L^2}$, and the energy inequality, which says that, in turn, since u is a solution of the equation, $\|\nabla u_\theta\|_{L^2}$ is controlled by $\|u_\theta\|_{L^2}$ for every truncation $u_\theta = (u - \theta)^+$. □

1.1.1 Sobolev and energy inequalities

We next recall the Sobolev and energy inequalities.

Sobolev inequality

If v is supported in B_1, then

$$\|v\|_{L^p(B_1)} \leq C\|\nabla v\|_{L^2(B_1)}$$

for some $p(N) > 2$.

If we are not too picky, we can prove it by representing

$$v(x_0) = \int_{B_1} \frac{\nabla v(x) \cdot (x_0 - x)}{|x - x_0|^n} \, dx = \nabla v * G.$$

Since G "almost" belongs to $L^{N/(N-1)}$, any $p < 2N/(N-2)$ would do. The case $p = 2N/(N-2)$ requires another proof.

Energy inequality

If $u \geq 0$, $D_i a_{ij} D_j u \geq 0$, and $\varphi \in C_0^\infty(B_1)$, then

$$\int_{B_1} (\nabla[\varphi u])^2 \leq C \sup |\nabla \varphi|^2 \int_{B_1 \cap \text{supp}\,\varphi} u^2.$$

(Note that there is a loss going from one term to the other: $\nabla(\varphi u)$ versus u.)

We denote by Λ^α the term $\Delta^\alpha \theta = -(-\Delta)^\alpha \theta$.

Proof. We multiply $Lu = D_i(a_{ij}D_j(-u))$ by $\varphi^2 u$. Since everything is positive, we get

$$-\int \nabla^T(\varphi^2 u)A\nabla u \geq 0, \quad \text{where } A = (a_{ij}).$$

We have to transfer a φ from the left ∇ to the right ∇. For this, we use that

$$\int \nabla^T(\varphi u)Au(\nabla\varphi) \leq \varepsilon \int \nabla^T(\varphi u)A\nabla(\varphi u) + \frac{1}{\varepsilon}\int |\nabla\varphi|^2 u^2 \|A\|.$$

(Try it!)   □

1.1.2   Proof of the De Giorgi theorem

The proof of the De Giorgi theorem is now split into two parts:

- Step 1: From L^2 to L^∞.
- Step 2: Oscillation decay.

We start with Step 1.

Lemma 2 (From an L^2 to an L^∞ bound). *If $\|u^+\|_{L^2(B_1)}$ is small enough, namely smaller than $\delta_0(n, \lambda, \Lambda)$, then $\sup_{B_{1/2}} u^+ \leq 1$.*

Before going into the proof of this lemma, let us give a simple, geometric analogy that avoids some of the technicalities of the proof.

Suppose that Ω is a domain in $\mathbb{R}^n$ and $\partial\Omega$ is a minimal surface when restricted to B_1, in the sense that the boundary of any perturbation inside B_1 will have larger area.

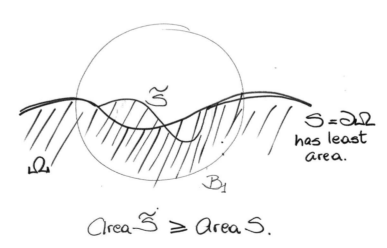

We want to prove that a minimal surface "has no cusps".

Lemma 3. *If* $\mathrm{Vol}(\Omega \cap B_1) \leq \varepsilon_0$*, a small enough constant, then* $\mathrm{Vol}(\Omega \cap B_{1/2}) = 0$*.*

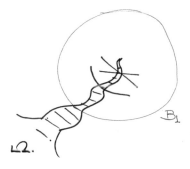

For this purpose, we will take dyadic balls $B_{r_k} = B_{\frac{1}{2}+2^{-k}}$ converging to $B_{1/2}$, and rings $R_{r_k} = B_{r_{k-1}} - B_{r_k}$; and we will find a *nonlinear* recurrence relation for $V_k = \mathrm{Vol}(\Omega \cap B_{r_k})$ for k even, that will imply that V_k goes to zero. In particular, Ω never reaches $B_{1/2}$. In this analogy, volume replaces the square of the L^2 norm of u, area the energy $\int |\nabla u|^2$, the isoperimetric inequality the Sobolev inequality, and the minimality of the area the energy inequality. The argument is based on the interplay between area and volume, as follows.

A_r controls V_k for $r \geq r_k$ in a nonhomogeneous way

We have

$$V_k \leq V_r \leq (\text{isoperimetric inequality}) \leq [\mathrm{Area}(\partial ``V_r")]^{N/N-1}$$
$$= (\text{two parts}) = [A_r + \mathrm{Area}(\partial\Omega \cap B_r)]^{N/N-1} \leq (2A_r)^{N/N-1}.$$

By minimality,

$$\mathrm{Area}(\partial\Omega \cap B_r) \leq A_r.$$

This is the "energy inequality".

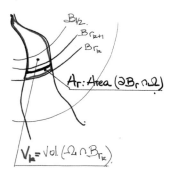

V_k controls A_r for some r in R_{k+1} in a homogeneous way

We have

$$\text{Vol}(\text{``}V_k\text{''} \setminus \text{``}V_{k+1}\text{''}) \sim \int_{r_{k+1}}^{r_k} A_r \geq 2^{-k} \inf_{r_{k+1} \leq r \leq r_k} A_r \ .$$

If we combine both estimates, we get (notice the different exponents)

$$V_{k+1} \leq \underset{r_{k+1} \leq r \leq r_k}{(2A_r)^{N/(N-1)}} \leq 2^{(N/(N-1))k} \, V_k^{\boxed{N/(N-1)}} \, !!\, .$$

If $V_0 < \varepsilon_0$, the build up in the exponent as we iterate beats the large geometric coefficient in the recurrence relation above, and V_k goes to zero. In particular, $B_{1/2}$ is "clean".

We now pass to the proof of Lemma 2. The origin becomes now plus infinity, $\|u\|_{L^2}$ plays the role of volume, and $\|\nabla u\|_{L^2}$ that of area. We have the added complication of having to truncate in space.

Proof of Lemma 2. We will consider a sequence of truncations $\varphi_k u_k$, where φ_k is a sequence of shrinking cut-off functions converging to $\chi_{B_{1/2}}$. More precisely:

$$\varphi_k \equiv \begin{cases} 1 & \text{for } |x| \leq 1 + 2^{-(k+1)} \\ 0 & \text{for } |x| \geq 1 + 2^{-k} \end{cases}$$

$$|\nabla \varphi_k| \leq C \, 2^k.$$

Note that $\varphi_k \equiv 1$ on $\text{supp}\,\varphi_{k+1}$,

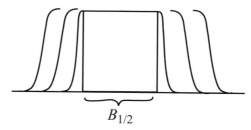

$$B_{1/2}$$

while u_k is a sequence of monotone truncations converging to $(u-1)^+$:

$$u_k = [u(1 - 2^{-k})]^+.$$

Note that, where $u_{k+1} > 0$, $u_k > 2^{-(k+1)}$. Therefore $\{(\varphi_{k+1}u_{k+1}) > 0\}$ is contained in $\{(\varphi_k u_k) > 2^{-(k+1)}\}$.

We will now show that, if $\|u\|_{L^2(B_1)} = A_0$ is small enough, then

$$A_k = \int (\varphi_k u_k)^2 \to 0.$$

In particular, $(u-1)^+\big|_{B_{1/2}} = 0$ a.e., that is, u never goes above 1 in $B_{1/2}$.

This is done, as in the example, through a (*nonlinear*) recurrence relation for A_k.

By the Sobolev inequality, we have

$$\left[\int (\varphi_{k+1} u_{k+1})^p\right]^{2/p} \leq C \int (\nabla[\varphi_{k+1} u_{k+1}])^2.$$

But, from Hölder,

$$\int (\varphi_{k+1} u_{k+1})^2 \leq \left[\int (\varphi_{k+1} u_{k+1})^p\right]^{2/p} \cdot |\{\varphi_{k+1} u_{k+1} > 0\}|^\varepsilon,$$

so we get

$$A_{k+1} \leq C \int [\nabla(\varphi_{k+1} u_{k+1})]^2 \cdot |\{\varphi_{k+1} u_{k+1} > 0\}|^\varepsilon.$$

We now control the right-hand side by A_k through the energy inequality. From energy we get

$$\int |\nabla(\varphi_{k+1} u_{k+1})|^2 \leq C\, 2^{2k} \int_{\mathrm{supp}\,\varphi_{k+1}} u_{k+1}^2$$

(but $\varphi_k \equiv 1$ on $\mathrm{supp}\,\varphi_{k+1}$)

$$\leq C\, 2^{2k} \int (\varphi_k u_k)^2 = C\, 2^{2k} A_k.$$

To control the last term, from the observation above:

$$|\{\varphi_{k+1} u_{k+1} > 0\}|^\varepsilon \leq |\{\varphi_k u_k > 2^{-k}\}|^\varepsilon,$$

and, by Chebyshev,

$$\leq 2^{2k\varepsilon} \left(\int (\varphi_k u_k)^2\right)^\varepsilon.$$

So we get

$$A_{k+1} \leq C\, 2^{4k} (A_k)^{1+\varepsilon}.$$

Then, for $A_0 = \delta$ small enough, $A_k \to 0$ (prove it). The build up of the exponent in A_k forces A_k to go to zero. In fact, A_k has faster than geometric decay, i.e., for any $M > 0$, $A_k < M^{-k}$ if $A_0(M)$ is small enough. $\square$

Corollary 4. *If u is a solution of $Lu = 0$ in B_1, then*

$$\|u\|_{L^\infty(B_{1/2})} \leq C\|u\|_{L^2(B_1)}.$$

Step 2: Oscillation decay

Let $\operatorname{osc}_D u = \sup_D u - \inf_D u$. We want to show:

Theorem 5. *If u is a solution of $Lu = 0$ in B_1, then there exists $\sigma(\lambda, \Lambda, n) < 1$ such that*

$$\operatorname{osc}_{B_{1/2}} u \le \sigma \operatorname{osc}_{B_1} u.$$

Let us begin with the following lemma.

Lemma 6. *Let $0 \le v \le 1$, $Lv \ge 0$ in B_1. Assume that $|B_{1/2} \cap \{v = 0\}| = \mu$, where $\mu > 0$. Then $\sup_{B_{1/4}} v \le 1 - \sigma(\mu)$.*

In other words, if v^+ is a subsolution of Lv, smaller than one in B_1, and is "far from 1" in a set of non-trivial measure, it cannot get too close to 1 in $B_{1/2}$.

The proof of Theorem 5 is based on the following idea. Suppose that, in B_1, $|u| \le 1$, i.e., $\operatorname{osc} u \le 2$. Then u is positive or negative at least half of the time. Say it is negative, i.e.,

$$|\{u^+ = 0\}| \ge \tfrac{1}{2}|B_1|.$$

Then, on $B_{1/2}$, u should not be able to be too close to 1. For u harmonic, for instance, this just follows from the mean value theorem.

To start with our proof of Lemma 6, we first observe that if

$$|\{u^+ = 0\}| \ge \left(1 - \frac{\delta}{2}\right)|B_1|,$$

then

$$\|u^+\|_{L^2(B_1)}^2 \le \delta/2,$$

and the previous lemma would tell us that $u^+|_{B_{1/2}} \le 1/2$.

So we must bridge the gap between knowing that $|\{u^+ = 0\}| \ge \tfrac{1}{2}|B_1|$ and knowing that $|\{u^+ = 0\}| \ge (1 - \frac{\delta}{2})B_1$.

A main tool is the De Giorgi isoperimetric inequality, which establishes that a function u with finite Dirichlet energy needs "some room in between" to go from a value (say 0) to another (say 1).

It may be considered a quantitative version of the fact that a function with a jump discontinuity cannot be in H^1.

Sublemma. *Let $0 \le w \le 1$. Set*

$$|A| = |\{w = 0\} \cap B_{1/2}|,$$
$$|C| = |\{w = 1\} \cap B_{1/2}|,$$
$$|D| = |\{0 < w < 1\} \cap B_{1/2}|.$$

Then, if $\int |\nabla w|^2 \le C_0^0$, we have

$$C_0|D| \ge C_1(|A|\,|C|^{1 - \frac{1}{n}})^2.$$

Proof. For x_0 in C we reconstruct w by integrating along any of the rays that go from x_0 to a point in A:

$$1 = w(x_0) = \int w_r \, dr, \quad \text{or}$$

$$|A| \le \text{area of } S(A) \le \int_D \frac{|\nabla w(y)| \, dy}{|x_0 - y|^{n-1}}$$

$$\left(w_r \, dr \, d\sigma \le \frac{|\nabla w| r^{n-1} dr \, d\sigma}{r^{n-1}} \right).$$

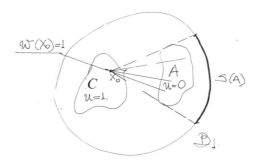

Integrating x_0 on C,

$$|A| \, |C| \le \int_D |\nabla w(y)| \left(\int_C \frac{dx_0}{|x_0 - y|^{n-1}} \right) dy.$$

Among all C with the same measure $|C|$, the integral in x_0 is maximized by the ball of radius $|C|^{1/n}$, centered at y:

$$\int_C \cdots \le |C|^{1/n}.$$

Hence,

$$|A| \, |C| \le |C|^{1/n} \left(\int_D |\nabla w|^2 \right)^{1/2} |D|^{1/2}.$$

Since $\int |\nabla w|^2 \le C_0^0$, the proof is complete. $\qquad \square$

With this sublemma, we go to the proof of Lemma 6.

Idea of the proof of Lemma 6. We will consider a dyadic sequence of truncations approaching 1,

$$v_k = [v - (1 - 2^{-k})]^+,$$

and their renormalizations

$$w_k = 2^k v_k.$$

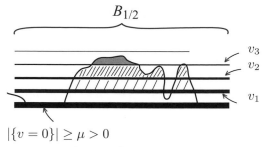

$$|\{v = 0\}| \geq \mu > 0$$

From the isoperimetric inequality, each time we truncate we expect the measure of the support to decay in a quantitative way. After a finite number of steps, the measure of the support of w_{k_0} will fall below the critical value $\delta/2$, and w_k will only be able to reach halfway towards 1, i.e.,

$$v|_{B_{1/2}} \leq 1 - 2^{-k_0}.$$

We will be interested in the set $C_k = \{v_k > 2^{-(k+1)}\} = \{w_k > 1/2\}$, its complement $A_k = \{v_k = 0\}$, and the transition set $D_k = [C_k - C_{k-1}]$.

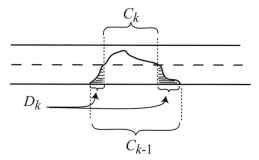

We will show that, by applying the isoperimetric inequality and the previous lemma in a finite number of steps $k_0 = k_0(\lambda, \Lambda, \mu)$,

$$|C_{k_0}| = 0.$$

Then $\sigma(\mu) = 2^{-k_0}$. Note that:

(a) $A_0 = \mu$ ($\mu = 1/2$ will do for our case).

(b) By the energy inequality, since $|w_k|_{B_1} \leq 1$,

$$\int_{B_{1/2}} |\nabla w_k|^2 \leq C.$$

(c) If C_k gets small enough,

$$4 \int (w_k)^2 \leq |C_k| < \delta,$$

we apply Lemma 6 to $2w_k$ and $2w_k|_{B_{1/4}} \leq 1$, and we are done.

We iterate this argument with $2 \min(w_k, \frac{1}{2}) = w$. If $|C_k|$ stays bigger than δ after a finite number of steps $k_0 = k(\delta, \mu)$, we get $\sum |D_k| \geq |B_{1/2}|$. This is impossible, so for some $k < k_0$, $|C_k| \leq \delta$, which makes $|C_{k+1}| = 0$ from the first part of the proof. $\qquad\square$

Corollary 7. $\operatorname{osc}_{B_{2^{-k}}} u \leq \lambda^k \operatorname{osc}_{B_1} u.$

Corollary 8. $u \in C^\alpha(B_{1/2})$ *with* $\lambda = 2^{-\alpha}$ *(defining α).*

Corollary 9. *If* $\|u\|_{L^\infty(R^n)} \leq C$, *then u is constant.*

The argument in Lemma 2 is very useful (and powerful) when two quantities of different homogeneity compete with each other: area and volume (in a minimal surface), or area and harmonic measure, or harmonic measure and volume as in free boundary problems.

1.2 Integral diffusion and the quasi-geostrophic equation

Nonlinear evolution equations with integral diffusions arise in many contexts: In turbulence [13], in boundary control problems [8], in problems of planar crack propagation in 3-D, in surface flame propagation, in "mean field games" theory, in mathematical finance [10] and in the quasi-geostrophic equation [6].

1.2.1 Quasi-geostrophic flow equation

The quasi-geostrophic (Q-G) equation is a 2-D "Navier–Stokes type" equation. In 2-D, Navier–Stokes simplifies considerably, since

(a) incompressibility (div $\vec{v} = 0$) implies that $(-v_2, v_1)$ is a gradient:

$$(-v_2, v_1) = \nabla\varphi;$$

(b) curl $\vec{v}$ is a scalar, $\theta = \operatorname{curl} \vec{v} = \Delta\varphi.$

The Navier–Stokes equation thus becomes a system:

$$\begin{cases} \theta_t + \vec{v}\,\nabla\theta = \Delta\theta, \\ \operatorname{curl} \vec{v} = \theta. \end{cases}$$

For the Q-G equation, we still have that

$$(-v_2, v_1) = \nabla\varphi.$$

But the potential φ is related to vorticity by $\theta = (-\Delta)^{1/2}\varphi$. That is, the final system becomes (we denote $\Delta^\alpha\theta = -(-\Delta)^\alpha\theta$)

$$\theta_t + \vec{v}\,\nabla\theta = \Delta^{1/2}\theta,$$

and $(-v_2, v_1) = (R_1\theta, R_2\theta)$, where R_i are the Riesz transforms of θ.

Note that we are in a critical case, since the regularization term $(\Delta^{1/2}\theta)$ is of the same order as the transport term $(\vec{v}\nabla\theta)$.

1.2.2   Riesz transforms and the dependence of v on θ

More precisely, we can deduce this relation through Fourier transform:

$$\hat{\theta} = |\zeta|\hat{\varphi} \qquad \text{and} \qquad \hat{v} = (-i\zeta_2\hat{\varphi}, \, i\zeta_1\hat{\varphi}).$$

In particular,

$$\hat{v} = \left(-\frac{i\zeta_2}{|\zeta|}\hat{\theta}, \, \frac{i\zeta_1}{|\zeta|}\hat{\theta}\right).$$

The multipliers $i\xi_j/|\xi|$ are classical operators, called *Riesz transforms*, that correspond in physical space x to convolution with kernels

$$R_j(x) = \frac{x_j}{|x|^{n+1}},$$

i.e.,

$$v_i^j(x) = \int R_j(x-y)\theta(y)\,dy.$$

Note that, on one hand,

$$\|v\|_{L^2(R^n)} = \|\hat{v}\|_{L^2(R^n)} \leq \|\hat{\theta}\|_{L^2(R^n)},$$

that is, the Riesz transforms are bounded operators from L^2 to L^2. On the other hand, R is neither integrable at zero nor at infinity. It is a remarkable theorem that, because of the spherical cancellation on R (mean value zero and smoothness), we have the following: The operator $R * \theta = v$ is a bounded operator from L^p to L^p for any $1 < p < \infty$ (Calderón–Zygmund). Unfortunately, it is easy to show that *singular* integral operators are not bounded from L^∞ to L^∞. They are bounded, though, from BMO to BMO.

1.2.3   BMO spaces

What is BMO? It is the space of functions with *bounded mean oscillation*. That is, in any cube Q the "average of u minus its average" is bounded by a constant C,

$$\frac{1}{|Q|}\int_Q \left|u(x) - \frac{1}{Q}\int_Q u(y)\,dy\right|dx \leq C.$$

The smallest C good for all cubes defines a seminorm (as it does not distinguish a constant that we may factor out). The space of functions u in BMO of the unit cube is smaller than any L^p $(p < \infty)$ but not included in L^∞ (for this, $(\log|x|)^-$ is a typical example).

In fact, functions u in BMO have "exponential" integrability,

$$\int_{Q_1} e^{C|u|} < \infty.$$

1.2.4 The fractional Laplacian and harmonic extensions

The fractional Laplacian $\Delta^\alpha \theta$ can be defined as convolution with a singular kernel $(0 < \alpha < 1)$,

$$\Delta^\alpha \theta(x_0) = C(\alpha) \text{ p.v.} \int \frac{[\theta(x) - \theta(x_0)]}{|x - x_0|^{n+2\alpha}} \, dx,$$

or through Fourier transform

$$\widehat{(-\Delta)^\alpha \theta}(\xi) = |\xi|^{2\alpha} \hat{\theta}(\xi).$$

Note that the kernel

$$K = C(\alpha)|x|^{-(n+2\alpha)}$$

is singular near zero, so, in principle, some cancellation in u is expected for the integral to converge. For instance, θ bounded and in C^2 near x_0 suffices. Also, $C(\alpha) \sim (1 - \alpha)$ guarantees that, as $\alpha \to 1$, $\Delta^\alpha \theta$ converges to $\Delta\theta$.

A particularly interesting case is the case $\alpha = 1/2$, since in this case $(-\Delta)^{1/2} u$ coincides with the Dirichlet to Neumann map. More precisely, given θ defined for x in $\mathbb{R}^n$, we extend it to θ^* defined for (x, y) in $(\mathbb{R}^{n+1})^+$ by combining it with the Poisson kernel:

$$P_y(x) = \frac{C \, y}{(y^2 + |x|^2)^{\frac{n+1}{2}}} = y^{-n} P_1(x/y).$$

Then $\theta^*(x, y)$ satisfies

$$\Delta_{x,y} \theta^* = 0 \quad \text{in} \quad \mathbb{R}^n \times \mathbb{R}^+$$

and it can be checked that $\Lambda^{1/2}\theta(x_0) = D_y\theta^*(x_0, 0)$ in two ways:

(a) Represent $\theta^*(x_0, h)$ as

$$\theta^*(x_0, h) = [P_h * \theta](x_0)$$

and take the limit on the difference quotient

$$D_y\theta^*(x_0, 0) = \lim_{h \to 0} \frac{\theta^*(x_0, h) - \theta^*(x_0, 0)}{h},$$

or

(b) Fourier-transform in x:

$$\widehat{\theta^*}(\xi, y) \quad \text{satisfies} \quad |\xi|^2 \widehat{\theta^*} = D_{yy}\widehat{\theta^*}.$$

Thus

$$\widehat{\theta^*}(\xi, y) = \widehat{\theta}(\xi)e^{-y|\xi|}.$$

In particular,

$$D_y\widehat{\theta}(\xi, 0) = -\widehat{\theta}(\xi)|\xi| = \widehat{(\Delta^{1/2}\theta)}(\xi).$$

Hence, we can make sense of the Green's and "energy" formula for the half Laplacian: Let $\sigma(x), \theta(x)$ be two "nice, decaying" functions defined in $\mathbb{R}^n$, and $\bar{\sigma}(x, y)$, $\bar{\theta}(x, y)$ decaying extensions into $(\mathbb{R}^{n+1})^+$. Then we have

$$\int_{\mathbb{R}^n} \sigma(\bar{\theta})_\nu = \int_{(\mathbb{R}^{n+1})^+} \nabla_{x,y}\bar{\sigma}\nabla_{(x,y)}\bar{\theta} + \int_{(\mathbb{R}^{n+1})^+} \bar{\sigma}\Delta_{x,y}\bar{\theta}.$$

If we choose $\bar{\theta}(x, y)$ and the harmonic extension θ^*, the term $\bar{\theta}_\nu(x, 0)$ becomes $-\Delta^{1/2}\theta$, and $\Delta\theta^* \equiv 0$, giving us

$$\int_{\mathbb{R}^n} \sigma(-\Delta^{1/2})\theta = \int_{(\mathbb{R}^{n+1})^+} \nabla\sigma\nabla\theta^*.$$

Further, if we choose

$$\sigma = (\theta - \lambda)^+ \quad \text{and} \quad \bar{\sigma} = (\theta^* - \lambda)^+$$

(i.e., the *truncation* of the *extension* of θ), we get

$$\int_{\mathbb{R}^n} (\theta - \lambda)^+(-\Lambda^{1/2}\theta) = \int_{(\mathbb{R}^{n+1})^+} [\nabla(\theta^* - \lambda)^+]^2 \, dx \, dy.$$

To complete our discussion, we point out that the harmonic extension θ^* of θ is the one extension that minimizes Dirichlet energy,

$$E(\theta^*) = \int_{\mathbb{R}^{n+1}_+} |\nabla\theta^*|^2,$$

and that this minimum defines the $H^{1/2}$ norm of θ. In particular, we obtain

$$\int_{\mathbb{R}^n} (\theta - \lambda)^+(-\Delta^{1/2})\theta = \iint [\nabla(\theta^* - \lambda)^+]^2 \, dx \, dy$$

$$\geq \iint [\nabla(\theta - \lambda)^*_+]^2 \, dx \, dy = \|(\theta - \lambda)^+\|^2_{H^{1/2}}$$

since *the harmonic extension of the truncation has less energy than the truncation of the harmonic extension.*

To recapitulate

The operator $\Delta^{1/2}$ is interesting because:

(a) It can be understood as a "surface diffusion" process.

(b) It is the "Euler–Lagrange equation" of the $H^{1/2}$ energy.

(c) Being of "order 1", diffusion competes with transport.

In fact, the derivatives of θ are

$$D_{X_j}\theta = R_j(\Delta^{1/2}\theta) \quad \text{and} \quad \Delta^{1/2}(\theta) = \sum_j R_j(D_{X_j}\theta),$$

where R_j, the Riesz transform, is the singular integral operator with symbol $i\xi_j/|\xi|$.

1.2.5 Regularity

The regularity theory for the quasi-geostrophic equation is based on two linear transport regularity theorems: Theorems 10 and 11.

Theorem 10. *Let θ be a (weak) solution of*

$$\theta_t + v\Delta\theta = (\Delta^{1/2})\theta \quad in \quad \mathbb{R}^n \times [0,\infty)$$

for some incompressible vector field v (with no a priori bounds) and initial data θ_0 in L^2. Then

$$\|\theta(\,\cdot\,,1)\|_{L^\infty(\mathbb{R}^n)} \leq C\|\theta(\,\cdot\,,0)\|_{L^2(\mathbb{R}^n)}.$$

Remarks. (i) All we ask from v is that the energy inequality makes sense for any function $h(\theta)$ with linear growth. Formally, if we multiply and integrate, we may write

$$\int_{T_1}^{T_2}\int_{\mathbb{R}^n} h(\theta)v\nabla\theta = \iint v\nabla H(\theta) = \iint \operatorname{div} v H(\theta) = 0,$$

where $H'(\theta) = h(\theta)$.

Therefore, the contribution of the transport term in the energy inequality vanishes. In the case of the Q-G equation, this can be attained by rigorously constructing θ in a particular way, for instance as a limit of solutions in increasing balls B_K.

(ii) From the scaling of the equation: For any λ,

$$\theta_\lambda = \frac{1}{\lambda}\theta(\lambda x, \lambda t)$$

is again a solution (with a different v). From Theorem 10 we obtain

$$\|\theta(\,\cdot\,,t_0)\|_{L_x^\infty} = t_0\|\theta_{t_0}(\,\cdot\,,1)\|_{L^\infty} \leq t_0\|\theta_{t_0}(\,\cdot\,,0)\|_{L^2} = t_0^{-n/2}\|\theta_0\|_{L^2}.$$

That is, uniform decay for large times.

The proof of Theorem 10 is a baby version of the De Giorgi theorem based on the interplay between the energy inequality (that controls the derivatives of θ by θ itself), and the Sobolev inequality (that controls θ by its derivatives). It is a baby version because, as in the minimal surface example, no cut-off in space is necessary.

The energy inequality is attained, as usual, by multiplying the equation with a truncation of θ,

$$(\theta_\lambda) = (\theta - \lambda)^+,$$

and integrating in $\mathbb{R}^n \times [T_1, T_2]$.

As we pointed out before, the term corresponding to transport vanishes, and we get

$$\frac{1}{2} \int \left[(\theta_\lambda)^2(y, T_2) - (\theta_\lambda)^2(y, T_1) \right] dy + 0 = \iint_{\mathbb{R}^n \times [T_1, T_2]} \theta_\lambda \Lambda^{1/2} \theta \, dy \, dt.$$

The last term corresponds, for the harmonic extension $\theta^*(x, z)$ to ($x \in \mathbb{R}^n$, $z \in \mathbb{R}^+$), to

$$\int_{T_1}^{T_2} dt \left(\int_{\mathbb{R}^n} (\theta^*)_\lambda(y, 0, t) D_z(\theta^*)(y, 0, t) \, dy \right)$$

$$= -\int_{T_1}^{T_2} \iint_{\mathbb{R}^{n+1}_+} \nabla(\theta^*)_\lambda(y, z, t) \nabla \theta^*(y, z, t) \, dy \, dz$$

$$= -\int_{T_1}^{T_2} dt \iint_{\mathbb{R}^{n+1}_+} [\nabla \theta_\lambda^*]^2 \, dy \, dz.$$

Note that $(\theta^*)_\lambda$ is *not* the harmonic extension of θ_λ, but the truncation of the extension of θ, i.e., $(\theta^* - \lambda)^+$.

Nevertheless, it is an admissible extension of θ_λ (going to zero at infinity) and, as such,

$$\|\theta_\lambda^*\|_{H^1(\mathbb{R}^{n+1}_+)} \geq \|\theta_\lambda\|_{H^{1/2}(\mathbb{R}^n)}.$$

Therefore we end up with the following energy inequality:

$$\|\theta_\lambda(\,\cdot\,, T_2)\|_{L^2}^2 + \int_{T_1}^{T_2} \|\theta_\lambda\|_{H^{1/2}}^2 \, dt \leq \|\theta_\lambda(T_1)\|_{L^2}^2.$$

We will denote by $A = (A_{T_1, T_2})$ the term on the left and by $B = B_{T_1}$ the one on the right. Therefore, B_{T_1} controls in particular (from the Sobolev inequality) all of the future:

$$\sup_{t \geq T_1} \|\theta(t)\|_{L^2}^2 + \int_{T_1}^{\infty} \|\theta(t)\|_{L^p}^2 \leq B_{T_1}.$$

This combination, in turn, actually controls

$$\|\theta\|_{L^q(\mathbb{R}^n \times [T_1, \infty))}^2$$

for some q with $2 < q < p$, in the following way. Every such q is a convex combination

$$q = \alpha 2 + (1 - \alpha)p = \frac{1}{r}2 + \frac{1}{s}p$$

for r, s appropriate conjugate exponents.

Therefore, fixing such a q, we have, for each time t,

$$\int \theta^q \leq \left(\int \theta^2\right)^{1/r} \cdot \left(\int \theta^p\right)^{1/s}.$$

We choose $s = p/2 \ (> 1)$ and integrate in t. For the corresponding q, we get

$$\|\theta\|_{L^q(\mathbb{R}^n \times [T_1, \infty))}^q \leq \sup_{t \geq t_1} \|\theta\|_{L^2(\mathbb{R}^n)}^{2/r} \cdot \int_{T_1}^{\infty} \|\theta\|_{L^p}^2 \leq (B_{T_1})^{1+\frac{1}{r}} = (B_{T_1})^{q/2}.$$

We call $C_{T_1} = \|\theta\|_{L^q(\mathbb{R}^n \times [T_1, \infty])}^{2/q}$, i.e., $C_{T_1} \leq B_{T_1}$.

We are ready to prove the L^∞ bound. For that purpose, we will find a recurrence relation for the constants

$$C_{T_k}(\theta_k)$$

of a sequence of increasing cut-offs $\lambda_k = 1 - 2^{-k}$ of θ (i.e., $\theta_k = \theta_{\lambda_k}$) and cut-offs in time $T_k = 1 - 2^{-k}$, that will imply that $\theta_\infty = (\theta - 1)^+ \equiv 0$ for $t > 1$.

Indeed, on one hand, from Sobolev:

$$C_{T_k}(\theta_k) \leq B_{T_k}(\theta_k).$$

We now invert the relation. For $I = [T_{k-1}, T_k] \times \mathbb{R}^n$, we have

$$\iint_I (\theta_k)^2 \leq \left[\iint_I \theta_k^q\right]^{2/q} |\{\theta_k > 0\} \cap I|^{1/\bar{q}} = \alpha \cdot \beta$$

(by Hölder with θ^2 and $\chi_{\theta_k > 0}$) with $\bar{q}$ the conjugate exponent to $q/2$.

In turn, $\alpha \leq C_{T_{k-1}}(\theta_{k-1})$ and, by going from k to $k-1$, we can estimate (this should sound familiar by now):

$$\beta = |\{\theta_{k-1} > 2^{-k}\} \cap I|^{1/\bar{q}} \leq \left[2^{qk} \iint_I (\theta_{k-1})^q\right]^{1/\bar{q}} \quad \text{(by Chebyshev)},$$

since $\theta_k \leq \theta_{k-1}$ and further $\theta_k > 0$ implies $\theta_{k-1} > 2^{-k}$.

That is, $\beta \leq 2^{Ck} \left[C_{T_{k-1}}(\theta_{k-1})\right]^\varepsilon$ and, putting together the estimates for α and β,

$$\iint_I (\theta_k)^2 \leq 2^{Ck} \left[C_{T_{k-1}}(\theta_{k-1})\right]^{1+\varepsilon}.$$

But then,

$$\inf_{[T_{k-1} < t < T_k]} B_t(\theta_k) \leq 2^k 2^{Ck} \cdot \left[C_{T_{k-1}}(\theta_{k-1})\right]^{1+\varepsilon}.$$

We obtain the recurrence relation

$$C_{T_k}(\theta_k) \leq 2^{\bar{C}k} \left[C_{T_{k-1}}(\theta_{k-1}) \right]^{1+\varepsilon}.$$

Due to the $1 + \varepsilon$ nonlinearity, $C_{T_k}(\theta_k) \to 0$ if $C_0(\theta^+)$ was small enough, i.e., if $\|\theta_0\|_{L^2} \leq \delta_0$, then $\|\theta(\,\cdot\,,t)\|_{L^\infty} \leq 1$, for $t \geq 1$. Since the equation is linear in θ, we can apply this result to $(\delta_0/\|\theta_0\|_{L^2})\theta$, which gives

$$\|\theta(\,\cdot\,,t)\|_{L^\infty} \leq \frac{\|\theta_0\|_{L^2}}{\delta_0} \quad \text{for } t \geq 1.$$

We now pass to the issue of regularity, i.e., the "oscillation lemma". Having shown boundedness for the Q-G equation, our situation is now the following. We have a solution θ that satisfies the energy bound:

$$\begin{cases} \sup_t \|\theta(t)\|^2_{L^2(\mathbb{R}^n)} + \|D_{1/2}\theta\|^2_{\mathbb{R}^{n+1}_+} \leq C \\[2mm] \text{and also} \ \ \|\theta\|_{L^\infty(X,t)} \leq 1, \end{cases} \tag{\#}$$

and we want to prove that θ is Hölder continuous. To do this, we need to reproduce the local in space De Giorgi method. Of the velocity field, we may assume now (being the Riesz transform of θ) that

$$\text{div } v = 0, \qquad \sup_t \left(\|v\|^2_{L^2(\mathbb{R}^n)} + \|v\|_{\mathrm{BMO}(\mathbb{R}^n)} \right) \leq C. \tag{$*$}$$

We decouple v from θ, and will prove a linear theorem, where for v satisfying $(*)$ and θ satisfying $(\#)$ and the equation

$$\theta_t + v\nabla\theta = \Delta^{1/2}\theta,$$

we have:

Theorem 11. *θ is locally C^α.*

To simplify the notation, we will assume that θ exists for $t \geq -4$ and will focus on the point $(X,t) = (0,0)$. The Hölder continuity will be proven through an oscillation lemma, i.e., we will prove that on a geometric sequence of cylinders

$$\Gamma_k = B_{4^{-k}} \times [4^{-k}, 0]$$

the oscillation of θ,

$$\omega_k = \sup_{\Gamma_k} \theta - \inf_{\Gamma_k} \theta,$$

decreases geometrically, i.e.,

$$\omega_{k+1} \leq \mu\omega_k \quad \text{for } \mu < 1.$$

This is proved in several steps, following the L^2 to L^∞ and oscillation lemmas discussed before.

The underlying idea is the following: Suppose that, on the cylinder $\Gamma_0 = B_1 \times [-1,0]$, θ lies between -1 and 1. Then at least half of the time it will be below or above zero. Let us say that it is below zero. Then, because of the diffusion process, by the time we are at the top of the cylinder and near zero, θ should have gone uniformly strictly below 1, so now $-1 \leq \theta \leq 1 - \delta$ and the oscillation ω has been reduced.

If we achieve this result, we renormalize and repeat. How do we achieve this oscillation reduction? For the heat equation, this will just follow from simple properties of the fundamental solution.

Here, following De Giorgi, we proceed in two steps. First, we show that if θ is "most of the time negative" or very tiny in $B_1 \times [-1,0]$, then, indeed, it cannot stick to the value 1 close to the top of the cylinder and so it goes strictly below 1 in, say, $B_{1/4} \times [-1/4, 0]$.

Next we have to close the gap between "being negative most of the time" and "being negative half of the time", since this last statement is what we can verify at each step.

This takes a finite sequence of cut-offs and renormalizations, exploiting the fact that for θ to go from a level (say 0) to another (say 1), some minimal amount of energy is necessary (the De Giorgi isoperimetric inequality). Finally, once this has been reached, we can iterate.

In our case, the arguments are complicated by the global character of the diffusion that may cancel the local effect that we described above. Luckily, we may encode the global effect locally into the harmonic extension, but this requires some careful treatment.

The first technical complication is that we must now truncate not only in θ and t but also in X, yet this does not have the effect of fully localizing the energy inequality, as a global term remains.

In the light of the iterative interaction between the Sobolev and energy inequalities, let us explore a little bit what kind of energy formulas we may expect after a cut-off in space.

Let us start with a cut-off in x and z, for $\theta(x,t)$ and its harmonic extension $\theta^*(x,z,t)$. That is, η is a smooth nonnegative function of x, z with support in $B_4^* = B_4 \times (-4, 4)$, and as usual we multiply the equation by $\eta^2 \theta_\lambda^*$ (which coincides with θ_λ for $z = 0$) and integrate.

We get the following terms:

$$2 \int_{T_1}^{T_2} \int \eta^2 \theta_\lambda \theta_t \, dx \, dt \equiv \int \eta^2 (\theta_\lambda)^2 (T_2) \, dx - \int \eta^2 (\theta_\lambda)^2 (T_1) \, dx. \qquad (I)$$

Next we have the transport term, an extra term not usually present in the energy inequality:

$$2 \iint \eta^2 \theta_\lambda v \nabla \theta \, dx \, dt = \iint \eta^2 \operatorname{div}[v(\theta_\lambda)^2] \, dx \, dt$$
$$= -\iint 2\eta \nabla \eta \, [v(\theta_\lambda)^2] \, dx \, dt. \tag{II}$$

We split the term into the two factors $(\nabla \eta) v \, \theta_\lambda$ and $\eta \theta_\lambda$, the logic being that v is almost bounded and thus the first term is almost like the standard right-hand side in the energy inequality while the second would be absorbed by the energy.

For each fixed t, we get

$$|\mathrm{II}| \leq \int_{T_1}^{T_2} \|\eta \theta_\lambda\|_{L^{2n/(n-1)}} \|\nabla \eta [v \theta_\lambda]\|_{L^{2n/(n+1)}}$$
$$\leq \int_{T_1}^{T_2} \varepsilon \|\eta \theta_\lambda\|_{L^{2n/(n-1)}}^2 + \frac{1}{\varepsilon} \|\nabla \eta [v \theta_\lambda]\|_{L^{2n/(n+1)}}^2 .$$

But $2n/(n+1) < 2$, so we can split by Hölder $[\nabla \eta] \theta_\lambda$ in L^2 and v in a (large) L^p, more precisely L^{2n} since we have that v is in every L^p.

That is,

$$\mathrm{II} \leq \varepsilon \int_{T_1}^{T_2} \|\eta \theta_\lambda\|_{L^{2n/(n-1)}}^2 + \frac{1}{\varepsilon} \int_{T_1}^{T_2} \|v\|_{L^{2n}(B_2)}^2 \|[\nabla \eta] \theta_\lambda\|_{L^2}^2 .$$

(Remember that, by hypothesis, $\|v\|_{L^{2n}(B_2)} \leq C$ for every t.)

Hence,

$$\mathrm{II} \leq \varepsilon \int_{T_1}^{T_2} \|\eta \theta_\lambda\|_{L^{2n/(n-1)}}^2 + \frac{1}{\varepsilon} C \int_{T_1}^{T_2} \|[\nabla \eta] \theta_\lambda\|_{L^2}^2 .$$

Finally, III is our energy term, i.e.,

$$\mathrm{III} = \iint \eta^2 \theta_\lambda \Delta^{1/2} \theta .$$

Using the harmonic extension θ^*, we get that

$$\mathrm{III} = -\iint \eta^2 \theta_\lambda \theta_\nu^* \, dx \, dt = -\iiint \nabla_{x,z} (\eta^2 \theta_\lambda^*) \nabla \theta^* \, dx \, dz \, dt .$$

By the standard energy inequality computation, we get that

$$\mathrm{III} \leq -\iiint [\nabla_{x,z} \eta \theta_\lambda^*]^2 \, dx \, dz \, dt + \iiint (\nabla \eta)^2 (\theta_\lambda^*)^2 .$$

We may choose η to be a cut-off in x and z or to integrate to infinity in z, if we have control of θ_λ^* in z.

Remark. If, for some reason, we know that $(\theta_\lambda^*) \equiv 0$ in $B_1 \times \{z_0\}$ for some z_0, then we may cut off only in x and still stop the integration at z_0.

Putting together I, II and III, we get

$$\sup_{T_1 \leq t \leq T_2} \|\eta\theta_\lambda\|_{L^2}^2 + \int_{T_1}^{T_2} \|\nabla(\eta\theta_\lambda^*)\|_{L^2}^2$$

$$\leq \|\eta\theta_\lambda(T_1)\|_{L^2}^2 + \varepsilon \int_{T_1}^{T_2} \|\eta\theta_\lambda\|_{L^{2n/(n-1)}}^2$$

$$+ \frac{1}{\varepsilon} C \int_{T_1}^{T_2} \|(\nabla\eta)\theta_\lambda\|_{L^2}^2 + \int_{T_1}^{T_2} \|(\nabla\eta)\theta_\lambda^*\|_{L^2}^2 .$$

Notice that $\eta\theta_\lambda^*$ is one extension of $\eta\theta_\lambda$ and therefore the term in the left

$$\int_{T_1}^{T_2} \|\nabla(\eta\theta_\lambda^*)\|_{L^2}^2 \quad \text{controls} \quad \int_{T_1}^{T_2} \|\eta\theta_\lambda\|_{H^{1/2}}^2,$$

so the left-hand side controls, by the Sobolev inequality, the term

$$\int_{T_1}^{T_2} \|\eta\theta_\lambda\|_{L^{2n/(n-1)}}^2$$

and absorbs the ε term on the right.

We finally get the energy estimate:

$$\sup_{T_1 \leq t \leq T_2} \|\eta\theta_\lambda\|_{L^2}^2 + \int_{T_1}^{T_2} \|\nabla(\eta\theta_\lambda^*)\|_{L^2}^2$$

$$\leq \|\eta\theta_\lambda(T_1)\|_{L^2}^2 + \int_{T_1}^{T_2} \|(\nabla\eta)\theta_\lambda^*\|_{L^2_{x,z}}^2 + \phi \int_{T_1}^{T_2} \|(\nabla\eta)\theta_\lambda\|_{L^2}^2 \tag{1.2.1}$$

where the constant ϕ depends only on the BMO seminorm and mean value of the velocity v (namely the constant in $(*)$).

In fact, if it weren't because of the term on the right

$$\int_{T_1}^{T_2} \|(\nabla\eta)\theta_\lambda^*\|_{L^2_{x,z}}^2$$

involving the extra variable z, everything would reduce to $\mathbb{R}^n$, and we would have the usual interplay between the Sobolev and energy inequalities, as in the global case, and a straightforward adaptation of the second-order case would work.

In light of this obstruction, let us reassess the situation. As we mentioned before, our first lemma (Lemma 12) would be (following the iterative scheme) to show that if, say, the L^2 norm of θ^+ is very small (and $\theta^+ \leq 2$) in $B_4 \times [-4,0]$, then θ^+ is strictly less than 2 in $B_1 \times [-1,0]$, i.e., $\theta^+ \leq \gamma_0 < 2$ for some γ_0.

Let us see how this can work.

1.2.6　A geometric description of the argument

Starting up, for a fixed time t, we decompose θ^* into two parts:

- θ_1^* goes to zero linearly as $z \to 0$ for $|x| < 1/2$.

- θ_2^* has a very small trace in L^2, so it becomes very small in L^∞ as z grows.

Given δ, we may assume that $\theta^*(X, \delta) \leq C\delta$ since we can choose $\|\theta\|_{L^2}$ as small as we please.

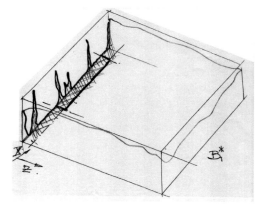

Therefore, the first truncation $\theta_{\lambda_0}^*$ is controlled by its trace, with very small L^2 norm, and the very narrow sides (of size δ), whose influence decays exponentially moving inwards in X:

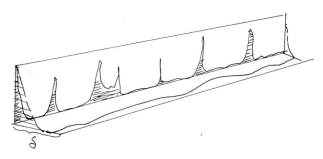

We will try now to perpetuate, in our inductive scheme, this configuration.

The idea of the inductive scheme is then as follows:

(1) In X, we will cut dyadically (as in De Giorgi) converging to $\chi_{B_{1/2}}$.

(2) In θ, also dyadically converging to λ (where $1 < \lambda < 2$).

(3) In Z, though, we will cut at a very fast geometric rate, going to zero (as δ^k).

The reason why we may hope to maintain this configuration is because inherent to the De Giorgi argument is the very fast decay (faster than geometric decay) of the L^2 norm of the truncation θ_k.

The idea is that, on one hand, the fast cut-off in Z will make the influence of the tiny sides decay so much in X that at the level of the next cut-off (in X) it will be wiped out by the dyadic cut-off in θ, while the contribution of the trace θ_k will decay so fast (faster than M^{-k} if we choose c_0 very small) that

$$\theta_k^*(X, \delta^k) \leq \theta_k * P_{\delta^k} \leq \|\theta_k\|_{L^2} \|P_{\delta^k}\|_{L^2}$$

will also be wiped out by the consecutive truncation.

At the end of the process, at time t_0, we have only information on the trace θ, but we can go inwards by harmonicity to complete the proof.

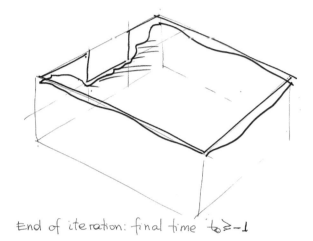

End of iteration: final time $t_0 \geq -1$

1.2.7 First part

In this first part, we prove that a solution θ between 0 and 2, with very small L^2 norm, separates from $\theta = 2$ in a smaller cylinder.

Lemma 12. *We assume that*

$$\|v\|_{L^\infty(-4, 0; \, \text{BMO}(\mathbb{R}^N))} + \sup_{-4 \leq t \leq 0} \left| \int_{B_4} v(t, x) \, dx \right| \leq C_0. \tag{1.2.2}$$

Then there exist $\varepsilon_0 > 0$ and $\lambda > 0$ such that, for every solution θ to (1.2.1), the following property holds true: If we have $\theta^* \leq 2$ in $[-4, 0] \times B_4^*$ and

$$\int_{-4}^0 \int_{B_4^*} (\theta^*)_+^2 \, dx \, dz \, ds + \int_{-4}^0 \int_{B_4} (\theta)_+^2 \, dx \, ds \leq \varepsilon_0,$$

then $(\theta)_+ \leq 2 - \lambda$ on $[-1, 0] \times B_1$.

Proof. The proof follows the strategy discussed above. First, we introduce some previous tools. Since the method was based on the control of θ^* by two harmonic functions, one for the local data and the other for the far away, before starting the proof we build two useful barriers.

Step 1: Barriers

The first barrier is the following:

Barrier $b_1(x,z)$

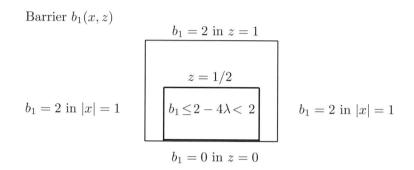

$b_1 = 2$ in $z = 1$

$z = 1/2$

$b_1 = 2$ in $|x| = 1$ $b_1 \leq 2 - 4\lambda < 2$ $b_1 = 2$ in $|x| = 1$

$b_1 = 0$ in $z = 0$

Then b_1 has the following properties:

(i) b_1 is harmonic in B_1^*;

(ii) $b_1 = 2$ in ∂B_1^* except $z = 0$;

(iii) $b_1 = 0$ in $\partial B_1^* \cap \{z = 0\}$;

and, for some $0 < \lambda$, we have $b_1 \leq 2 - 4\lambda < 2$ in $B_{1/2}^*$.

     The barrier b_1 implies that

θ_1 has linear decay as $y \to 0$.

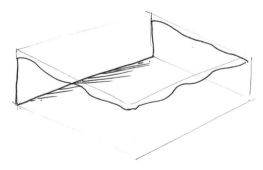

The barrier b_2 satisfies the following:

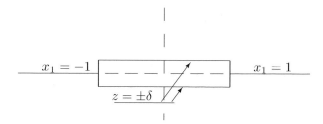

(i) b_2 is harmonic in D;

(ii) $b_2 = 0$ for $z \pm \delta$;

(iii) $b_2 = 1$ for $x_1 = 1$ and $b_2 = 0$ for $x_1 = -1$.

Then $b_2 \leq \bar{C} \cos(z/\delta)\, e^{-(1-x_1)/\delta}$. In particular, if $1 - x_1 = h \gg \delta$, we have $b_2 \leq \bar{C} e^{-h/\delta}$ and $\bar{C} = (\cos 1)^{-1}$.

Remark. Exponential decay also holds for $D_i a_{ij} D_j$ by applying the Harnack inequality to the intervals $I_k = \{k \leq x_1 \leq k + 1\}$.

Now we are ready to set the main inductive steps as discussed above. When we do so, we will realize that we have to start the process for some advanced value k of the step. So we will go back and do a first large step to cover the starting of the process.

Step 2: Setting of the constants

We recall that $\lambda > 0$ is defined by the fact that the barrier function satisfies that $b_1 < 2 - 4\lambda$ in $B^*_{1/2}$. Next, $\bar{C} = (\cos 1)^{-1}$ is the constant in the bound for the barrier function b_2. The smallness constant C_0 in the hypothesis of Lemma 12 will be chosen later as $C_0(\lambda, M)$.

We need to fix constants M for the rate of decay of the L^2 norm of the truncation θ_k and δ for the rate of decay of the support in z of θ_k^*.

We require:

(i) $n\bar{C} e^{-(2\delta)^{-k}} \leq \lambda 2^{-k-2}$ (with δ small so the side contribution is absorbed by the cut-off);

(ii) $\delta^n (M\delta^n)^{-k} \|P(1)\|_{L^2} \leq \lambda 2^{-k-2}$ (with $M(\delta)$ large to keep the support of the truncation in the δ^k strip);

(iii) $M^{-k} \geq C_0^k M^{-(k-3)(n+1/n)}$ for $k \geq 12n$ (so that the inductive decay gives us the fast geometric decay).

Here $P(1)$ denotes the restriction of the Poisson kernel $P(x, z)$ to $z \equiv 1$.

The choice is straightforward. We first construct δ to verify the first inequality in the following way. If $\delta < 1/4$, the inequality is true for $k > k_0$ due to the exponential decay. If necessary, we then choose δ smaller to make the inequality also valid for $k < k_0$. Now that δ has been fixed, we have to choose M large to satisfy the remaining inequalities. Note that the second inequality is equivalent to

$$\left(\frac{2}{\delta^n M}\right)^k \leq \frac{\lambda \delta^n}{4\|P(1)\|_{L^2}}.$$

It is sufficient to take

$$M \geq \sup\left(\frac{2}{\delta^n}, \frac{8\|P(1)\|_{L^2}}{\lambda \delta^n}\right).$$

The third inequality is equivalent to

$$\left(\frac{M}{C_0^N}\right)^{k/N} \geq M^{3(1+1/N)}.$$

For this it is enough to take $M \geq \sup(1, C_0^{2N})$. Indeed, this ensures $M^2/C_0^{2N} \geq M$ and hence

$$\left(\frac{M}{C_0^N}\right)^{k/N} \geq M^{k/(2N)} \geq M^6.$$

The main inductive step will be the following.

Step 3: Induction

We set

$$\theta_k = (\theta - C_k)_+,$$

with $C_k = 2 - \lambda(1 + 2^{-k})$. We consider a cut-off function in x only such that

$$\mathbf{1}_{\{B_{1+2^{-k-1}}\}} \leq \eta_k \leq \mathbf{1}_{\{B_{1+2^{-k}}\}}, \qquad |\nabla \eta_k| \leq C 2^k,$$

and we let

$$A_k = 2\int_{-1-2^{-k}}^{0}\int_{0}^{\delta^k}\!\!\int_{\mathbb{R}^N} |\nabla(\eta_k \theta_k^*)|^2 \, dx \, dz \, dt + \sup_{[-1-2^{-k},1]}\int_{\mathbb{R}^N}(\eta_k \theta_k)^2 \, dx \, dt.$$

We want to prove that, for every $k \geq 0$,

$$A_k \leq M^{-k}, \text{and} \tag{1.2.3}$$

$$\eta_k \theta_k^* \text{ is supported in } 0 \leq z \leq \delta^k \tag{1.2.4}$$

(it vanishes at δ^k, and thus we can extend it by zero).

Step 4: Starting the process

We prove in this step that, if ε_0 is small enough, then (1.2.3) is verified for $0 \leq k \leq 12N$, and that (1.2.4) is verified for $k = 0$. We use the energy inequality (1.2.1) with cut-off function $\eta_k(x)\psi(z)$ where ψ is a fixed cut-off function in z only. Taking the mean value of (1.2.1) in T_1 between -4 and -2, we find that (1.2.3) is verified for $0 \le k \le 12N$ if ε_0 is taken such that

$$C2^{24N}(1+\phi)\varepsilon_0 \le M^{-12N}. \tag{1.2.5}$$

We have used that $|\nabla \eta_k|^2 \le C'2^{24N}$ for $0 \le k \le 12N$. Let us consider now the support property (1.2.4). By the maximum principle, we have

$$\theta^w \le (\theta_+ \mathbf{1}_{B_4}) * P(z) + b_1(x,z)$$

in $\mathbb{R}^+ \times B_4^*$, where $P(z)$ is the Poisson kernel. Indeed, the right-hand side function is harmonic, positive, and the trace on the boundary is bigger than the one of θ^*.

From Step 1 we have $b_1(x,z) \le 2 - 4\lambda$. Moreover,

$$\|\theta_+ \mathbf{1}_{B_4} * P(z)\|_{L^\infty(z \ge 1)} \le C\|P(1)\|_{L^2}\sqrt{\varepsilon_0} \le C\sqrt{\varepsilon_0}.$$

Choosing ε_0 small enough so that this constant is smaller than 2λ gives

$$\theta^* \le 2 - 2\lambda \quad \text{for } 1 \le z \le 2, \, t \ge 0, \, x \in B_2,$$

so

$$\theta_0^* = (\theta^* - (2 - 2\lambda))_+ \le 0 \quad \text{for } 1 \le z \le 2, \, t \ge 0, \, x \in B_2.$$

Hence, $\eta_0 \theta_0^*$ vanishes for $1 = \delta^0 \le z \le 2$.

Step 5: Propagation of the support property (1.2.4)

Assume that (1.2.3) and (1.2.4) are verified at k. We want to show that (1.2.4) is verified at $(k+1)$. We will also show that the following is verified at k:

$$\eta_{k+1}\theta_{k+1}^* \le [(\eta_k\theta_k) * P(z)]\eta_{k+1} \quad \text{on } \overline{B}_k^*, \tag{1.2.6}$$

where $\overline{B}_k^* = B_{1+2^{-k}} \times [0, \delta^k]$. We want to control θ_k^* on this set by harmonic functions taking into account the contributions of the sides one by one. Consider $B_{1+2^{-k-1/2}} \times [0, \delta^k]$. On $z = \delta^k$ we have no contribution thanks to the induction property (1.2.4) at k (the trace is equal to 0). The contribution of the side $z = 0$ can be controlled by $\eta_k\theta_k * P(z)$. (It has the same trace as θ_k on $B_{1+2^{-k-1/2}}$.)

On each of the other sides we control the contribution by the following function of $x = (x_1, \ldots, x_N)$:

$$b_2((x_i - x^+)/\delta^k, z/\delta^k) + b_2((-x_i + x^-)/\delta^k, z/\delta^k),$$

where $x^+ = (1 + 2^{-k-1/2})$ and $x^- = -x^+$. Indeed, b_2 is harmonic, and on the side x_i^+ and x_i^- it is bigger than 2. Finally, by the maximum principle,

$$\theta_k^* \le \sum_{i=1}^{N} \left[b_2((x_i - x^+)/\delta^k, z/\delta^k) + b_2((-x_i + x^-)/\delta^k, z/\delta^k) \right] + (\eta_k\theta_k) * P(z).$$

From Step 1, for $x \in B_{1+2^{-k-1}}$:

$$\sum_{i=1}^{N} \left[b_2((x_i - x^+)/\delta^k, z/\delta^k) + b_2((-x_i + x^-)/\delta^k, z/\delta^k) \right]$$

$$\leq 2N\overline{C}e^{-\frac{2^{-k}}{4(\sqrt{2}+1)\delta^k}} \leq \lambda 2^{-k-2}$$

(thanks to Step 2). This gives (1.2.6), since

$$\theta_{k+1}^* \leq (\theta_k^* - \lambda 2^{-k-1})_+.$$

More precisely, this gives

$$\theta_{k+1}^* \leq ((\eta_k \theta_k) * P(z) - \lambda 2^{-k-2})_+.$$

So,

$$\eta_{k+1}\theta_{k+1}^* \leq ((\eta_k \theta_k) * P(z) - \lambda 2^{-k-2})_+.$$

From the second property of Step 2, we find that, for $\delta^{k+1} \leq z \leq \delta^k$,

$$|(\eta_k \theta_k) * P(z)| \leq \sqrt{A_k} \|P(z)\|_{L^2}$$

$$\leq \frac{M^{-k/2}}{\delta^{(k+1)N/2}} \|P(1)\|_{L^2} \leq \lambda 2^{-k-2}.$$

The last inequality makes use of Step 2. Therefore,

$$\eta_{k+1}\theta_{k+1}^* \leq 0 \quad \text{for } \delta^{k+1} \leq z \leq \delta^k.$$

Note, in particular, that with Step 4 this gives that (1.2.4) is verified up to $k = 12N + 1$ and (1.2.6) up to $k = 12N$.

Step 6: Propagation of property (1.2.3)

We show in this step that if (1.2.3) is true for $k-3$ and (1.2.4) is true for $k-3$, $k-2$ and $k-1$, then (1.2.3) is true for k.

First notice that, from Step 5, (1.2.4) is true at $k-2$, $k-1$, and k. We just need to show that

$$A_k \leq C_0^k (A_{k-3})^{1+1/N} \quad \text{for } k \geq 12N + 1, \tag{1.2.7}$$

with

$$C_0 = C\frac{2^{1+2/N}}{\lambda^{2/N}}. \tag{1.2.8}$$

Indeed, if we do the third inequality of Step 2, this will give us the result.

Step 7: Proof of (1.2.7)

Since $\eta_k\theta_k^*\mathbf{1}_{\{0<z<\delta^{k-1}\}}$ has the same trace at $z=0$ as $(\eta_k\theta_k)^*$ and the latter is harmonic, we have

$$\int_0^{\delta^{k-1}}\int_{\mathbb{R}^N}|\nabla(\eta_k\theta_k^*)|^2 = \int_0^\infty\int_{\mathbb{R}^N}|\nabla(\eta_k\theta_k^*\mathbf{1}_{\{0<z<\delta^{k-1}\}})|^2$$

$$\geq \int_0^\infty\int_{\mathbb{R}^N}|\nabla(\eta_k\theta_k)^*|^2$$

$$= \int_{\mathbb{R}^N}|\Lambda^{1/2}(\eta_k\theta_k)|^2.$$

Note that we have used (1.2.4) in the first equality. The Sobolev and Hölder inequalities give

$$A_{k-3} \geq C\|\eta_{k-3}\theta_{k-3}\|^2_{L^{2(N+1)/N}([-1-2^{-k-3},0]\times\mathbb{R}^N)}.$$

From (1.2.6),

$$\|\eta_{k-2}\theta_{k-2}^*\|^2_{L^{2(N+1)/N}} \leq \|P(1)\|^2_{L^1}\|\eta_{k-3}\theta_{k-3}\|^2_{L^{2(N+1)/N}}.$$

Hence,

$$A_{k-3} \geq C\|\eta_{k-2}\theta_{k-2}^*\|^2_{L^{2(N+1)/N}} + C\|\eta_{k-3}\theta_{k-3}\|^2_{L^{2(N+1)/N}}$$

$$\geq C\left(\|\eta_{k-2}\theta_{k-1}^*\|^2_{L^{2(N+1)/N}} + \|\eta_{k-2}\theta_{k-1}\|^2_{L^{2(N+1)/N}}\right).$$

Since η_k is a cut-off function in x, and using (1.2.4), we have that $\eta_k\theta_k^*$ vanishes on the boundary of $B_{1+2^{-k}}\times[-\delta^k,\delta^k]$. We can then apply the energy inequality (1.2.5) on $\eta_k\theta_k^*\mathbf{1}_{\{0<z<\delta^{k-1}\}}$. Taking the mean value of (1.2.5) in T_1 between $-1-2^{-k-1}$ and $-1-2^{-k}$, we find:

$$A_k \leq C2^{2k}(\phi+2)\left(\int_0^{\delta^k}\int_{\mathbb{R}^N}\eta_{k-1}^2\theta_k^2 + \int_0^{\delta^k}\int_{\mathbb{R}^N}\eta_{k-1}^2\theta_k^{*2}\right).$$

We have used here the fact that $|\nabla\eta|^2 \leq C2^{2k}\eta_{k-1}^2$. If $\theta_k>0$, then $\theta_{k-1}\geq 2^{-k}\lambda$. So,

$$\mathbf{1}_{\{\theta_k>0\}} \leq \frac{C2^k}{\lambda}\theta_{k-1},$$

and

$$\mathbf{1}_{\{\eta_{k-1}>0\}}\mathbf{1}_{\{\theta_k>0\}} \leq \frac{C2^k}{\lambda}\eta_{k-2}\theta_{k-1}.$$

Therefore,

$$\int\eta_{k-1}^2\theta_k^2 + \int\eta_{k-1}^2\theta_k^{*2}$$

$$\leq \frac{C2^{2k/N}}{\lambda^{2/N}}\left[\int(\eta_{k-2}\theta_{k-1})^{2(N+1)/N} + \int(\eta_{k-2}\theta_{k-1}^*)^{2(N+1)/N}\right],$$

and so

$$A_k \leq \frac{C 2^{k(2+2/N)}}{\lambda^{2/N}} A_{k-3}^{1+1/N}.$$

This gives (1.2.7), for C big enough compared to $\lambda^{2/N}$.   □

1.2.8   Second part

In the first part, we have established that, if $0 \leq \theta_+ \leq 2$ and its energy or norm is very small in B_4^*, then $\theta_+ \leq 2 - \lambda$ in B_j, i.e., the oscillation of θ actually decays.

We now want to get rid of the "very small" hypothesis.

This second lemma (Lemma 13) proves that, if $\theta_+ \leq 0$ "half of the time" and it only needs very little room (say, δ) to go from $\{\theta_+ \leq 0\}$ to $\{\theta \geq 1\}$, it is because $(\theta - 1)^+$ has already very small norm to start with. This produces a dichotomy: either the support of θ decreases substantially, or θ becomes small anyway, in the same spirit as De Giorgi's lemma.

Lemma 13. *For every $\varepsilon_1 > 0$ there exists a constant $\delta_1 > 0$ with the following property: For every solution θ to (1.2.1) with v verifying (1.2.2), if $\theta^* \leq 2$ in Q_4^* and*

$$|\{(x, z, t) \in Q_4^* \ : \ \theta^*(x, z, t) \leq 0\}| \geq \frac{|Q_4^*|}{2},$$

then the following implication holds true: If

$$|\{(x, z, t) \in Q_4^* \ : \ 0 < \theta^*(x, z, t) < 1\}| \leq \delta_1,$$

then

$$\int_{Q_1} (\theta - 1)_+^2 \, dx \, dt + \int_{Q_1^*} (\theta^* - 1)_+^2 \, dx \, dz \, dt \leq C \sqrt{\varepsilon_1}.$$

This lemma is, of course, the adapted version of the De Giorgi's isoperimetric inequality.

The idea of the proof is the following. We first throw away a small set of times for which $I_t = \int_{B_1^*} |\nabla u^*|^2 \, dx \, dz$ is very large:

$$I_t \geq \frac{K^2}{\varepsilon_1^2}.$$

This is a tiny set of times:

$$|S| \leq C \varepsilon^2 / k^2,$$

since

$$\iint |\nabla u^*|^2 \, dx \, dz \, dt \leq C_0.$$

Outside of S, for each time t, the isoperimetric inequality is valid:

$$|A| \, |B| \leq |D| \, K / \varepsilon_1.$$

But for some t, say $t < -\frac{1}{64}$, we may choose a slice where $|A| > \frac{1}{64}$ and $|D| \le \delta$. Then $|B| \le (64)^2 \delta \ K/\varepsilon_1 \le K\varepsilon_1$ if $\delta \sim \varepsilon^2$. In particular, $(\theta - 1)^+$ has very small L^2 norm for that t: $\|(\theta - 1)^+\|_{L^2} \le k\varepsilon_1$. But the energy inequality then controls the L^2 norm of $(\theta - 1)^+$ into the future. We now give a more detailed proof.

Proof. Take $\varepsilon_1 \ll 1$. From the energy inequality and using that $\theta^* \le 2$ in Q_4^*, we get

$$\int_{-4}^{0} \int_{B_1^*} |\nabla \theta_+^*|^2 \, dx \, dz \, dt \le C.$$

Let

$$K = \frac{4 \int |\nabla \theta_+^*|^2 \, dx \, dz \, dt}{\varepsilon_1}.$$

Then

$$\left| \left\{ t \ : \ \int_{B_1^*} |\nabla \theta_+^*|^2 (t) \, dx \, dz \ge K \right\} \right| \le \frac{\varepsilon_1}{4}. \tag{1.2.9}$$

For all $t \in \{t \ : \ \int_{B_1^*} |\nabla \theta_+^*|^2(t) \, dx \, dz \le K\}$, the isoperimetric inequality gives that

$$|\mathcal{A}(t)| |\mathcal{B}(t)| \le |\mathcal{C}(t)|^{1/2} K^{1/2},$$

where

$$\begin{aligned}
\mathcal{A}(t) &= \{(x, z) \in B_1^* \ : \ \theta^*(t, x, z) \le 0\}, \\
\mathcal{B}(t) &= \{(x, z) \in B_1^* \ : \ \theta^*(t, x, z) \ge 1\}, \\
\mathcal{C}(t) &= \{(x, z) \in B_1^* \ : \ 0 < \theta^*(t, x, z) < 1\}.
\end{aligned}$$

Let us set $\delta_1 = \varepsilon_1^8$ and

$$I = \left\{ t \in [-4, 0] \ : \ |\mathcal{C}(t)|^{1/2} \le \varepsilon_1^3 \ \text{ and } \ \int_{B_1^*} |\nabla \theta_+^*|^2(t) \, dx \, dz \le K \right\}.$$

First we have, using the Chebyshev inequality,

$$\left| \{ t \in [-4, 0] \ : \ |\mathcal{C}(t)|^{1/2} \ge \varepsilon_1^3 \} \right| \le \frac{|\{(t, x, z) \ : \ 0 < \theta^* < 1\}|}{\varepsilon_1^6}$$

$$\le \frac{\delta_1}{\varepsilon_1^6} \le \varepsilon_1^2 \le \varepsilon_1/4.$$

Hence $|[-4, 0] \setminus I| \le \varepsilon_1/2$. Secondly, we get for every $t \in I$ such that $|\mathcal{A}(t)| \ge 1/4$,

$$|\mathcal{B}(t)| \le \frac{|\mathcal{C}(t)|^{1/2} K^{1/2}}{|\mathcal{A}(t)|} \le 4C\varepsilon_1^{5/2} \le \varepsilon_1^2. \tag{1.2.10}$$

In particular,

$$\int_{B_1^*} \theta_+^{*\,2}(t)\,dx\,dz \le 4(|\mathcal{B}(t)| + |\mathcal{C}(t)|) \le 8\varepsilon_1^2.$$

But

$$\int_{B_1} \theta_+^2(t)\,dx = \int_{B_1} \theta_+^{*\,2}(t,x,z)\,dx - 2\int_0^z \int_{B_1} \theta_+^*(t)\partial_z\theta^*\,dx\,d\bar{z}$$

for any z. Therefore, integrating in z on $[0,1]$, we find:

$$\int_{B_1} \theta_+^2(t)\,dx \le \int_{B_1^*} \theta_+^{*\,2}(t,x,z)\,dx\,dz + 2\sqrt{K}\sqrt{\int_{B_1^*} \theta_+^{*\,2}(t)\,dx\,dz} \le C\sqrt{\varepsilon_1}.$$

We want to show that $|\mathcal{A}(t)| \ge 1/4$ for every $t \in I \cap [-1,0]$. First, since

$$|\{(t,x,z) \ : \ \theta^* \le 0\}| \ge |Q_4^*|/2,$$

there exists $t_0 \le -1$ such that $|\mathcal{A}(t_0)| \ge 1/4$. For this t_0, $\int \theta_+^2(t_0)\,dx \le C\sqrt{\varepsilon_1}$. Using the energy inequality, for any $r > 0$ (where $\nabla\eta$ is of order $1/r$) we have, for every $t \ge t_0$,

$$\int_{B_1} \theta_+^2(t)\,dx \le \int_{B_1} \theta_+^2(t_0)\,dx + \frac{C(t-t_0)}{r} + Cr.$$

Let us choose r such that

$$Cr + C\sqrt{\varepsilon_1} \le 1/128.$$

So, for $t - t_0 \le \delta^* = r/(128C)$, we have

$$\int_{B_1} \theta_+^2(t)\,dx \le \frac{1}{64}.$$

(Note that δ^* does not depend on ε_1. Hence we can suppose $\varepsilon_1 \ll \delta^*$.) Moreover,

$$\theta_+^*(z) = \theta_+ + \int_0^z \partial_z\theta_+^*\,d\bar{z} \le \theta_+ + \sqrt{z}\left(\int_0^z |\partial_z\theta_+^*|^2\,d\bar{z}\right)^{1/2}.$$

So, for $t - t_0 \le \delta^*$, $t \in I$ and $z \le \varepsilon_1^2$, we have, for each x,

$$\theta_+^*(t,x,z) \le \theta_+(t,x) + \left(\varepsilon_1^2\int_0^\infty |\partial_z\theta_+^*|^2\,d\bar{z}\right)^{1/2}.$$

The integral, in x only, of the square of the right-hand side term is less than $1/8 + C\sqrt{\varepsilon_1} \le 1/4$. So, by Chebyshev, for every fixed $z \le \varepsilon_1$,

$$|\{x \in B_1, \ \theta_+^*(t,x,z) \ge 1\}| \le \frac{1}{4}.$$

Integrating in z on $[0, \varepsilon_1^2]$ gives

$$|\{z \leq \varepsilon_1^2, \ x \in B_1, \ \theta_+^*(t) \geq 1\}| \leq \frac{\varepsilon_1^2}{4}.$$

First we work in $B_1 \times [0, \varepsilon_1^2]$. Since $|\mathcal{C}(t)| \leq \varepsilon_1^6$, this gives

$$|\mathcal{A}(t)| \geq |B_1| \varepsilon_1^2 - |\{z \leq \varepsilon_1^2, \ x \in B_1, \ \theta_+^*(t) \geq 1\}| - |\mathcal{C}(t)|$$

$$\geq c_1^2(1 - 1/4) - c_1^6 \geq \varepsilon_1^2/2.$$

In the same way as in (1.2.10), we find that

$$|\mathcal{B}(t)| \leq \frac{|\mathcal{C}(t)|^{1/2} K^{1/2}}{|\mathcal{A}(t)|} \leq C\sqrt{\varepsilon_1},$$

and

$$|\mathcal{A}(t)| \geq 1 - |\mathcal{B}(t)| - |\mathcal{C}(t)| \geq 1 - 2\sqrt{\varepsilon_1} - \varepsilon_1^6 \geq 1/4.$$

Hence, for every $t \in [t_0, t_0 + \delta^*] \cap I$ we have $|\mathcal{A}(t)| \geq 1/4$. On $[t_0 + \delta^*/2, t_0 + \delta^*]$ there exists $t_1 \in I$ ($\delta^* \geq \varepsilon_1/4$). And so, we can construct an increasing sequence t_n with $0 \geq t_n \geq t_0 + n\delta^*/2$, such that $|\mathcal{A}(t)| \geq 1/4$ on $[t_n, t_n + \delta^*] \cap I \supset [t_n, t_{n+1}] \cap I$. Finally, on $I \cap [-1, 0]$ we have $|\mathcal{A}(t)| \geq 1/4$. This gives from (1.2.10) that, for every $t \in I \cap [-1, 0]$, $|\mathcal{B}(t)| \leq \varepsilon_1/16$. Hence,

$$|\{\theta^* \geq 1\}| \leq \varepsilon_1/16 + \varepsilon_1/2 \leq \varepsilon_1.$$

Since $(\theta^* - 1)_+ \leq 1$, this gives that

$$\int_{Q_1^*} (\theta^* - 1)_+^2 \, dx \, dz \, dt \leq \varepsilon_1.$$

We have, for every t, x fixed,

$$\theta - \theta^*(z) = -\int_0^z \partial_z \theta^* \, dz.$$

So,

$$(\theta - 1)_+^2 \leq 2\left((\theta^*(z) - 1)_+^2 + \left(\int_0^z |\nabla \theta^*| \, dz\right)^2\right)$$

for any z. Hence we have

$$(\theta - 1)_+^2 \leq \frac{2}{\sqrt{\varepsilon_1}} \int_0^{\sqrt{\varepsilon_1}} (\theta^* - 1)_+^2 \, dz + 2\sqrt{\varepsilon_1} \int_0^{\sqrt{\varepsilon_1}} |\nabla \theta^*|^2 \, dz.$$

Therefore,

$$\int_{Q_1} (\theta - 1)_+^2 \, dx \, ds \leq C\sqrt{\varepsilon_1}. \qquad \square$$

1.2.9   Oscillation lemma

We are now ready to iterate the process for the oscillation lemma.

Lemma 14. *There exists $\lambda^* > 0$ such that, for every solution θ to (1.2.3) with v satisfying (1.2.2), if $\theta^* \leq 2$ in Q_1^* and*

$$|\{(t, x, z) \in Q_1^* \; : \; \theta^* \leq 0\}| \geq \frac{1}{2},$$

then $\theta^* \leq 2 - \lambda^*$ in $Q_{1/16}^*$.

Note that λ^* depends only on N and C in (1.2.2).

Proof. For every $k \in \mathbb{N}$, $k \leq K_+ = E(1/\delta_1 + 1)$ (where δ_1 is defined in Lemma 13 for ε_1 such that $4C\sqrt{\varepsilon_1} \leq \varepsilon_0$, and ε_0 is defined in Lemma 12), we define

$$\overline{\theta}_k = 2(\overline{\theta}_{k-1} - 1) \quad \text{with } \overline{\theta}_0 = \theta.$$

So we have $\overline{\theta}_k = 2^k(\theta - 2) + 2$. Note that, for every k, $\overline{\theta}_k$ verifies the same equation, $\overline{\theta}_k \leq 2$, and $|\{(t, x, z) \in Q_1^* \; : \; \overline{\theta}_k \leq 0\}| \geq \frac{1}{2}$. Assume that, for all those k, $|\{0 < \overline{\theta}_k^* < 1\}| \geq \delta_1$. Then, for every k,

$$|\{\overline{\theta}_k^* < 0\}| = |\{\overline{\theta}_{k-1}^* < 1\}| \geq |\{\overline{\theta}_{k-1}^* < 0\}| + \delta_1.$$

Hence, $|\{\overline{\theta}_{K_+}^* \leq 0\}| \geq 1$ and $\overline{\theta}_{K_+}^* < 0$ almost everywhere, which means that $2^{K_+}(\theta^* - 2) + 2 < 0$, or $\theta^* < 2 - 2^{-K_+}$, and in this case we are done.

Else, there exists $0 \leq k_0 \leq K_+$ such that $|\{0 < \overline{\theta}_{k_0}^* < 1\}| \leq \delta_1$. From Lemma 13 and Lemma 12 (applied on $\overline{\theta}_{k_0+1}$) we get $(\overline{\theta}_{k_0+1})_+ \leq 2 - \lambda$, which means that

$$\theta \leq 2 - 2^{-(k_0+1)}\lambda \leq 2 - 2^{-K_+}\lambda$$

in $Q_{1/8}$. Consider the function b_3 defined by

(i) $\Delta b_3 = 0$ in $B_{1/8}^*$;

(ii) $b_3 = 2$ on the sides of the cube except for $z = 0$;

(iii) $b_3 = 2 - 2^{-K_+}\inf(\lambda, 1)$ on $z = 0$.

We have $b_3 < 2 - \lambda^*$ in $B_{1/16}^*$, and from the maximum principle we get $\theta^* \leq b_3$. $\square$

1.2.10   Proof of Theorem 11

We fix $t_0 > 0$ and consider $t \in [t_0, \infty) \times \mathbb{R}^N$. We define

$$F_0(s, y) = \theta(t + st_0/4, x + t_0/4(y - x_0(s))),$$

where $x_0(s)$ is a solution to

$$\begin{cases} \dot{x}_0(s) = \dfrac{1}{|B_4|} \displaystyle\int_{x_0(s)+B_4} v(t + st_0/4, x + yt_0/4) \, dy, \\ x_0(0) = 0. \end{cases}$$

Note that $x_0(s)$ is uniquely defined from the Cauchy–Lipschitz theorem. We set

$$\tilde{\theta}_0^*(s, y) = \frac{4}{\sup_{Q_4^*} F_0^* - \inf_{Q_4^*} F_0^*} \left(F_0^* - \frac{\sup_{Q_4^*} F_0^* + \inf_{Q_4^*} F_0^*}{2} \right),$$

$$v_0(s, y) = v(t + st_0/4, x + t_0/4(y - x_0(s))) - \dot{x}_0(s),$$

and then, for every $k > 0$,

$$F_k(s, y) = F_{k-1}(\tilde{\mu}s, \tilde{\mu}(y - x_k(s))),$$

$$\tilde{\theta}_k^*(s, y) = \frac{4}{\sup_{Q_4^*} F_k^* - \inf_{Q_4^*} F_k^*} \left(F_k^* - \frac{\sup_{Q_4^*} F_k^* + \inf_{Q_4^*} F_k^*}{2} \right),$$

$$\dot{x}_k(s) = \frac{1}{|B_4|} \int_{x_k(s)+B_4} v_{k-1}(\tilde{\mu}s, \tilde{\mu}y) \, dy,$$

$$x_k(0) = 0,$$

$$v_k(s, y) = v_{k-1}(\tilde{\mu}s, \tilde{\mu}(y - x_k(s))) - \dot{x}_k(s).$$

where $\tilde{\mu}$ will be chosen later. We divide the proof into several steps.

Step 1. For $k = 0$, $\tilde{\theta}_0$ is a solution to (1.2.3) in $[-4, 0] \times \mathbb{R}^N$, $\|v_0\|_{\text{BMO}} = \|v\|_{\text{BMO}}$, $\int v_0(s) \, dy = 0$ for every s, and $|\tilde{\theta}_0| \leq 2$. Assume that it is true at $k - 1$. Then

$$\partial_s F_k = \tilde{\mu} \partial_s \tilde{\theta}_{k-1} - \tilde{\mu} \dot{x}_k(s) \cdot \nabla \tilde{\theta}_{k-1}.$$

So $\tilde{\theta}_k$ is a solution to (1.2.3) and $|\tilde{\theta}_k| \leq 2$. By construction, for every s we have $\int_{B_4} v_k(s, y) \, dy = 0$ and $\|v_k\|_{\text{BMO}} = \|v_{k-1}\|_{\text{BMO}} = \|v\|_{\text{BMO}}$. Moreover, we have

$$|\dot{x}_k(s)| \leq \int_{B_4} v_{k-1}(\tilde{\mu}(y - x_k(s))) \, dy$$

$$\leq C \|v_{k-1}(\tilde{\mu}y)\|_{L^p}$$

$$\leq C\tilde{\mu}^{-N/p} \|v_{k-1}\|_{L^p}$$

$$\leq C_p \tilde{\mu}^{-N/p} \|v_{k-1}\|_{\text{BMO}}.$$

So, for $0 \leq s \leq 1$, $y \in B_4$ and $p > N$,

$$|\tilde{\mu}(y - x_k(s))| \leq 4\tilde{\mu}(1 + C_p\tilde{\mu}^{-N/p}) \leq C\tilde{\mu}^{1-N/p}.$$

For $\tilde{\mu}$ small enough, this is smaller than 1.

Step 2. For every k we can use the oscillation lemma. If $|\{\tilde{\theta}_k^* \le 0\}| \ge \frac{1}{2}|Q_4^*|$, then we have $\tilde{\theta}_k^* \le 2 - \lambda^*$. Else we have $|\{-\tilde{\theta}_k^* \le 0\}| \ge \frac{1}{2}|Q_4^*|$ and applying the oscillation lemma on $-\tilde{\theta}_k^*$ gives $\tilde{\theta}_k^* \ge -2 + \lambda^*$. In both cases, this gives

$$|\sup \tilde{\theta}_k^* - \inf \tilde{\theta}_k^*| \le 2 - \lambda^*,$$

and so

$$\left| \sup_{Q_1^*} F_k^* - \inf_{Q_1^*} F_k^* \right| \le (1 - \lambda^*/2)^k \left| \sup_{Q_1^*} F_0^* - \inf_{Q_1^*} F_0^* \right|.$$

Step 3. For $s \le \tilde{\mu}^{2n}$,

$$\sum_{k=0}^{n} \tilde{\mu}^{n-k} x_k(s) \le \tilde{\mu}^{2n} \sum_{k=0}^{n} \frac{\tilde{\mu}^{n-k}}{\tilde{\mu}^{-N/p}} \le \frac{\tilde{\mu}^n}{2},$$

for $\tilde{\mu}$ small enough. So

$$\left| \sup_{[-\tilde{\mu}^{2n},0] \times B_{\tilde{\mu}^n/2}^*} \theta^* - \inf_{[-\tilde{\mu}^{2n},0] \times B_{\tilde{\mu}^n/2}^*} \theta^* \right| \le (1 - \lambda^*/2)^n.$$

This gives that θ^* is C^α at $(t, x, 0)$, and so θ is C^α at (t, x). $\qquad\square$

Bibliography

[1] L. Caffarelli, R. Kohn, and L. Nirenberg, *Partial regularity of suitable weak solutions of the Navier–Stokes equations*, Comm. Pure Appl. Math. **35** (1982), no. 6, 771–831.

[2] L. Caffarelli and A. Vasseur, *Drift diffusion equations with fractional diffusion and the quasi-geostrophic equation*, Ann. of Math. **171** (2010), no. 3, 1903–1930.

[3] D. Chae, *On the regularity conditions for the dissipative quasi-geostrophic equations*, SIAM J. Math. Anal. **37** (2006), no. 5, 1649–1656 (electronic).

[4] D. Chae and J. Lee, *Global well-posedness in the super-critical dissipative quasi-geostrophic equations*, Commun. Math. Phys. **233** (2003), no. 2, 297–311.

[5] C.-C. Chen, R. M. Strain, H.-T. Yau, and T.-P. Tsai, *Lower bound on the blow-up rate of the axisymmetric Navier–Stokes equations*, Int. Math. Res. Notices **9** (2008), 31 pp.

[6] P. Constantin, *Euler equations, Navier–Stokes equations and turbulence*, in: Mathematical Foundation of Turbulent Viscous Flows, Lecture Notes in Math., vol. 1871, pp. 1–43, Springer, Berlin, 2006.

[7] E. De Giorgi, *Sulla differenziabilità e l'analiticità delle estremaili degli integrali multipli regolari*, Mem. Accad. Sci. Torino Cl. Sci. Fis. Mat. Natur. **3** (1957), 25–43.

[8] G. Duvaut and J.-L. Lions, *Les inéquations en mécanique et en physique*, Travaux et Recherches Mathématiques, vol. 21, Dunod, Paris, 1972.

[9] A. Kiselev, F. Nazarov, and A. Volberg, *Global well-posedness for the critical 2D dissipative quasi-geostrophic equation*, Invent. Math. **167** (2007), 445–453.

[10] P. Laurence and S. Salsa, *Regularity of the free boundary of an American option on several assets*, Comm. Pure Appl. Math. **62** (2009), no. 7, 969–994.

[11] G. Seregin and V. Sverak, *On Type I singularities of the local axi-symmetric solutions of the Navier–Stokes equations*, Comm. Partial Differential Equations **34** (2009), no. 2, 171–201.

[12] A. Vasseur, *A new proof of partial regularity of solutions to Navier–Stokes equations*, Nonlinear Differential Equations Appl. **14** (2007), no. 5-6, 753–785.

[13] G. M. Zaslavsky, M. Edleman, H. Weitzner, B. A. Carreras, G. McKee, R. Bravenec, and R. Fonck, Large-scale behavior of the tokamak density fluctuations, Phys. Plasmas **7** (2000), 3691–3698.

Chapter 2

Recent Results on the Periodic Lorentz Gas

François Golse

Introduction: from particle dynamics to kinetic models

The kinetic theory of gases was proposed by J. Clerk Maxwell [34, 35] and L. Boltzmann [5] in the second half of the XIXth century. Because the existence of atoms, on which kinetic theory rested, remained controversial for some time, it was not until many years later, in the XXth century, that the tools of kinetic theory became of common use in various branches of physics such as neutron transport, radiative transfer, plasma and semiconductor physics, etc.

Besides, the arguments which Maxwell and Boltzmann used in writing what is now known as the "Boltzmann collision integral" were far from rigorous — at least from the mathematical viewpoint. As a matter of fact, the Boltzmann equation itself was studied by some of the most distinguished mathematicians of the XXth century —such as Hilbert and Carleman— before there were any serious attempts at deriving this equation from first principles (i.e., molecular dynamics). Whether the Boltzmann equation itself was viewed as a fundamental equation of gas dynamics, or as some approximate equation valid in some well identified limit is not very clear in the first works on the subject —including Maxwell's and Boltzmann's.

It seems that the first systematic discussion of the validity of the Boltzmann equation viewed as some limit of molecular dynamics —i.e., the free motion of a large number of small balls subject to binary, short range interaction, for instance elastic collisions— goes back to the work of H. Grad [26]. In 1975, O. E. Lanford gave the first rigorous derivation [29] of the Boltzmann equation from molecular dynamics. His result proved the validity of the Boltzmann equation for a very short

39

time of the order of a fraction of the reciprocal collision frequency. (One should also mention an earlier, "formal derivation" by C. Cercignani [12] of the Boltzmann equation for a hard sphere gas, which considerably clarified the mathematical formulation of the problem.) Shortly after Lanford's derivation of the Boltzmann equation, R. Illner and M. Pulvirenti managed to extend the validity of his result for all positive times, for initial data corresponding with a very rarefied cloud of gas molecules [27].

An important assumption made in Boltzmann's attempt at justifying the equation bearing his name is the "Stosszahlansatz", to the effect that particle pairs just about to collide are uncorrelated. Lanford's argument indirectly established the validity of Boltzmann's assumption, at least on very short time intervals.

In applications of kinetic theory other than rarefied gas dynamics, one may face the situation where the analogue of the Boltzmann equation for monatomic gases is linear, instead of quadratic. The linear Boltzmann equation is encountered for instance in neutron transport, or in some models in radiative transfer. It usually describes a situation where particles interact with some background medium — such as neutrons with the atoms of some fissile material, or photons subject to scattering processes (Rayleigh or Thomson scattering) in a gas or a plasma.

In some situations leading to a linear Boltzmann equation, one has to think of two families of particles: the moving particles whose phase space density satisfies the linear Boltzmann equation, and the background medium that can be viewed as a family of fixed particles of a different type. For instance, one can think of the moving particles as being light particles, whereas the fixed particles can be viewed as infinitely heavier, and therefore unaffected by elastic collisions with the light particles. Before Lanford's fundamental paper, an important —unfortunately unpublished— preprint by G. Gallavotti [19] provided a rigorous derivation of the linear Boltzmann equation assuming that the background medium consists of fixed entities, like independent hard spheres whose centers are distributed in the Euclidean space under Poisson's law. Gallavotti's argument already possessed some of the most remarkable features in Lanford's proof, and therefore must be regarded as an essential step in the understanding of kinetic theory.

However, Boltzmann's Stosszahlansatz becomes questionable in this kind of situation involving light and heavy particles, as potential correlations among heavy particles may influence the light particle dynamics. Gallavotti's assumption of a background medium consisting of independent hard spheres excluded this possibility. Yet, strongly correlated background media are equally natural, and should also be considered.

The periodic Lorentz gas discussed in these notes is one example of this type of situation. Assuming that heavy particles are located at the vertices of some lattice in the Euclidean space clearly introduces about the maximum amount of correlation between these heavy particles. This periodicity assumption entails a dramatic change in the structure of the equation that one obtains under the same scaling limit that would otherwise lead to a linear Boltzmann equation.

Figure 2.1: Left: Paul Drude (1863–1906); right: Hendrik Antoon Lorentz (1853–1928)

Therefore, studying the periodic Lorentz gas can be viewed as one way of testing the limits of the classical concepts of the kinetic theory of gases.

Acknowledgements. Most of the material presented in these lectures is the result of collaboration with several authors: J. Bourgain, E. Caglioti, H. S. Dumas, L. Dumas and B. Wennberg, whom I wish to thank for sharing my interest for this problem. I am also grateful to C. Boldighrini and G. Gallavotti for illuminating discussions on this subject.

2.1 The Lorentz kinetic theory for electrons

In the early 1900's, P. Drude [16] and H. Lorentz [30] independently proposed to describe the motion of electrons in metals by the methods of kinetic theory. One should keep in mind that the kinetic theory of gases was by then a relatively new subject: the Boltzmann equation for monatomic gases appeared for the first time in the papers of J. Clerk Maxwell [35] and L. Boltzmann [5]. Likewise, the existence of electrons had been established shortly before, in 1897 by J. J. Thomson.

The basic assumptions made by H. Lorentz in his paper [30] can be summarized as follows.

First, the population of electrons is thought of as a gas of point particles described by its phase-space density $f \equiv f(t, x, v)$, that is, the density of electrons at the position x with velocity v at time t.

Electron-electron collisions are neglected in the physical regime considered in the Lorentz kinetic model —on the contrary, in the classical kinetic theory of gases, collisions between molecules are important as they account for momentum and heat transfer.

However, the Lorentz kinetic theory takes into account collisions between electrons and the surrounding metallic atoms. These collisions are viewed as simple, elastic hard sphere collisions.

Since electron-electron collisions are neglected in the Lorentz model, the equation governing the electron phase-space density f is linear. This is at variance with the classical Boltzmann equation, which is quadratic because only binary collisions involving pairs of molecules are considered in the kinetic theory of gases.

With the simple assumptions above, H. Lorentz arrived at the following equation for the phase-space density of electrons $f \equiv f(t, x, v)$:

$$(\partial_t + v \cdot \nabla_x + \tfrac{1}{m} F(t, x) \cdot \nabla_v) f(t, x, v) = N_{\mathrm{at}}\, r_{\mathrm{at}}^2\, |v|\, \mathcal{C}(f)(t, x, v).$$

In this equation, $\mathcal{C}$ is the Lorentz collision integral, which acts on the only variable v in the phase-space density f. In other words, for each continuous function $\phi \equiv \phi(v)$, one has

$$\mathcal{C}(\phi)(v) = \int_{\substack{|\omega|=1 \\ \omega \cdot v > 0}} \big(\phi(v - 2(v \cdot \omega)\omega) - \phi(v)\big) \cos(v, \omega)\, d\omega,$$

and the notation $\mathcal{C}(f)(t, x, v)$ designates $\mathcal{C}(f(t, x, \cdot))(v)$.

The other parameters involved in the Lorentz equation are the mass m of the electron, and N_{at}, r_{at} respectively the density and radius of metallic atoms. The vector field $F \equiv F(t, x)$ is the electric force. In the Lorentz model, the self-consistent electric force —i.e., the electric force created by the electrons themselves— is neglected, so that F takes into account only the effect of an applied electric field (if any). Roughly speaking, the self consistent electric field is linear in f, so that its contribution to the term $F \cdot \nabla_v f$ would be quadratic in f, as would be any collision integral accounting for electron-electron collisions. Therefore, neglecting electron-electron collisions and the self-consistent electric field are both in accordance with assuming that $f \ll 1$.

The line of reasoning used by H. Lorentz to arrive at the kinetic equations above is based on the postulate that the motion of electrons in a metal can be adequately represented by a simple mechanical model —a collisionless gas of point particles bouncing on a system of fixed, large spherical obstacles that represent the metallic atoms. Even with the considerable simplification in this model, the argument sketched in the article [30] is little more than a formal analogy with Boltzmann's derivation of the equation now bearing his name.

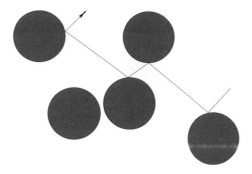

Figure 2.2: The Lorentz gas: a particle path

This suggests the mathematical problem of deriving the Lorentz kinetic equation from a microscopic, purely mechanical particle model. Thus, we consider a gas of point particles (the electrons) moving in a system of fixed spherical obstacles (the metallic atoms). We assume that collisions between the electrons and the metallic atoms are perfectly elastic, so that, upon colliding with an obstacle, each point particle is specularly reflected on the surface of that obstacle.

Undoubtedly, the most interesting part of the Lorentz kinetic equation is the collision integral which does not seem to involve F. Therefore we henceforth assume for the sake of simplicity that there is no applied electric field, so that

$$F(t, x) \equiv 0 \,.$$

In that case, electrons are not accelerated between successive collisions with the metallic atoms, so that the microscopic model to be considered is a simple, dispersing billiard system —also called a Sinai billiard. In that model, electrons are point particles moving at a constant speed along rectilinear trajectories in a system of fixed spherical obstacles, and specularly reflected at the surface of the obstacles.

More than 100 years have elapsed since this simple mechanical model was proposed by P. Drude and H. Lorentz, and today we know that the motion of electrons in a metal is a much more complicated physical phenomenon whose description involves quantum effects.

Yet the Lorentz gas is an important object of study in nonequilibrium satistical mechanics, and there is a very significant amount of literature on that topic —see for instance [44] and the references therein.

The first rigorous derivation of the Lorentz kinetic equation is due to G. Gallavotti [18, 19], who derived it from a billiard system consisting of randomly (Poisson) distributed obstacles, possibly overlapping, considered in some scaling limit —the Boltzmann–Grad limit, whose definition will be given (and discussed) below. Slightly more general, random distributions of obstacles were later considered by H. Spohn in [43].

While Gallavotti's theorem bears on the convergence of the mean electron density (averaging over obstacle configurations), C. Boldrighini, L. Bunimovich and Ya. Sinai [4] later succeeded in proving the almost sure convergence (i.e., for a.e. obstacle configuration) of the electron density to the solution of the Lorentz kinetic equation.

In any case, none of the results above says anything on the case of a periodic distribution of obstacles. As we shall see, the periodic case is of a completely different nature —and leads to a very different limiting equation, involving a phase-space different from the one considered by H. Lorentz, i.e., $\mathbf{R}^2 \times \mathbf{S}^1$, on which the Lorentz kinetic equation is posed.

The periodic Lorentz gas is at the origin of many challenging mathematical problems. For instance, in the late 1970s, L. Bunimovich and Ya. Sinai studied the periodic Lorentz gas in a scaling limit different from the Boltzmann–Grad limit studied in the present paper. In [7], they showed that the classical Brownian motion is the limiting dynamics of the Lorentz gas under that scaling assumption —their work was later extended with N. Chernov; see [8]. This result is indeed a major achievement in nonequilibrium statistical mechanics, as it provides an example of an irreversible dynamics (the heat equation associated with the classical Brownian motion) that is derived from a reversible one (the Lorentz gas dynamics).

2.2   The Lorentz gas in the Boltzmann–Grad limit with a Poisson distribution of obstacles

Before discussing the Boltzmann–Grad limit of the periodic Lorentz gas, we first give a brief description of Gallavotti's result [18, 19] for the case of a Poisson distribution of independent, and therefore possibly overlapping obstacles. As we shall see, Gallavotti's argument is in some sense fairly elementary, and yet brilliant.

First we define the notion of a Poisson distribution of obstacles. Henceforth, for the sake of simplicity, we assume a 2-dimensional setting.

The obstacles (metallic atoms) are disks of radius r in the Euclidean plane $\mathbf{R}^2$, centered at $c_1, c_2, \ldots, c_j, \ldots \in \mathbf{R}^2$. Henceforth, we denote by

$$\{c\} = \{c_1, c_2, \ldots, c_j, \ldots\} = \text{ a configuration of obstacle centers.}$$

We further assume that the configurations of obstacle centers $\{c\}$ are distributed under Poisson's law with parameter n, meaning that

$$\text{Prob}(\{\{c\} \mid \#(A \cap \{c\}) = p\}) = e^{-n|A|} \frac{(n|A|)^p}{p!},$$

where $|A|$ denotes the surface, i.e., the 2-dimensional Lebesgue measure of a measurable subset A of the Euclidean plane $\mathbf{R}^2$.

This prescription defines a probability on countable subsets of the Euclidean plane $\mathbf{R}^2$.

Obstacles may overlap: in other words, configurations $\{c\}$ such that

for some $j \neq k \in \{1, 2, \ldots\}$, one has $|c_i - c_j| < 2r$

are not excluded. Indeed, excluding overlapping obstacles means rejecting obstacle configurations $\{c\}$ such that $|c_i - c_j| \leq 2r$ for some $i, j \in \mathbf{N}$. In other words, $\mathrm{Prob}(d\{c\})$ is replaced with

$$\frac{1}{Z} \prod_{i > j \geq 0} \mathbf{1}_{|c_i - c_j| > 2r} \, \mathrm{Prob}(d\{c\}),$$

where $Z > 0$ is a normalizing coefficient. Since the term $\prod_{i > j \geq 0} \mathbf{1}_{|c_i - c_j| > 2r}$ is not of the form $\prod_{k \geq 0} \phi_k(c_k)$, the obstacles are no longer independent under this new probability measure.

Next we define the billiard flow in a given obstacle configuration $\{c\}$. This definition is self-evident, and we give it for the sake of completeness, as well as in order to introduce the notation.

Given a countable subset $\{c\}$ of the Euclidean plane $\mathbf{R}^2$, the billiard flow in the system of obstacles defined by $\{c\}$ is the family of mappings

$$(X(t; \cdot, \cdot, \{c\}), V(t; \cdot, \cdot, \{c\})) : Z_r \times \mathbf{S}^1 \to Z_r \times \mathbf{S}^1$$

where

$$Z_r := \{y \in \mathbf{R}^2 | \, \mathrm{dist}(x, c_j) > r \text{ for all } j \geq 1\},$$

defined by the following prescription.

Whenever the position X of a particle lies outside the surface of any obstacle, that particle moves at unit speed along a rectilinear path:

$$\dot{X}(t; x, v, \{c\}) = V(t; x, v, \{c\}),$$
$$\dot{V}(t; x, v, \{c\}) = 0, \qquad \text{whenever } |X(t; x, v, \{c\}) - c_i| > r \text{ for all } i,$$

and, in case of a collision with the i-th obstacle, is specularly reflected on the surface of that obstacle at the point of impingement, meaning that

$$X(t^+; x, v, \{c\}) = X(t^-; x, v, \{c\}) \in \partial B(c_i, r),$$
$$V(t^+; x, v, \{c\}) = \mathcal{R}\left[\frac{X(t; x, v, \{c\}) - c_i}{r}\right] V(t^-; x, v, \{c\}),$$

where $\mathcal{R}[\omega]$ denotes the reflection with respect to the line $(\mathbf{R}\omega)^\perp$:

$$\mathcal{R}[\omega]v = v - 2(\omega \cdot v)\omega, \quad |\omega| = 1.$$

Then, given an initial probability density $f_{\{c\}}^{\mathrm{in}} \equiv f_{\{c\}}^{\mathrm{in}}(x, v)$ on the single-particle phase-space with support outside the system of obstacles defined by $\{c\}$, we define its evolution under the billiard flow by the formula

$$f(t, x, v, \{c\}) = f_{\{c\}}^{\mathrm{in}}(X(-t; x, v, \{c\}), V(-t; x, v, \{c\})), \quad t \geq 0.$$

Let $\tau_1(x, v, \{c\}), \tau_2(x, v, \{c\}), \ldots, \tau_j(x, v, \{c\}), \ldots$ be the sequence of collision times for a particle starting from x in the direction $-v$ at $t = 0$ in the configuration of obstacles $\{c\}$. In other words,

$$\tau_j(x, v, \{c\}) = \sup\{t \mid \#\{s \in [0, t] \mid \operatorname{dist}(X(-s, x, v, \{c\}); \{c\}) = r\} = j - 1\}.$$

Letting $\tau_0 = 0$ and $\Delta\tau_k = \tau_k - \tau_{k-1}$, the evolved single-particle density f is a.e. defined by the formula

$$f(t, x, v, \{c\}) = f^{\mathrm{in}}(x - tv, v)\,\mathbf{1}_{t < \tau_1}$$

$$+ \sum_{j \geq 1} f^{\mathrm{in}}\left(x - \sum_{k=1}^{j} \Delta\tau_k V(-\tau_k^-) - (t - \tau_j)V(-\tau_j^+), V(-\tau_j^+)\right) \mathbf{1}_{\tau_j < t < \tau_{j+1}}.$$

In the case of physically admissible initial data, there should be no particle located inside an obstacle. Hence we assumed that $f^{\mathrm{in}}_{\{c\}} = 0$ in the union of all the disks of radius r centered at the $c_j \in \{c\}$. By construction, this condition is obviously preserved by the billiard flow, so that $f(t, x, v, \{c\})$ also vanishes whenever x belongs to a disk of radius r centered at any $c_j \in \{c\}$.

As we shall see shortly, when dealing with bounded initial data, this constraint disappears in the (yet undefined) Boltzmann–Grad limit, as the volume fraction occupied by the obstacles vanishes in that limit.

Therefore, we shall henceforth neglect this difficulty and proceed as if f^{in} were any bounded probability density on $\mathbf{R}^2 \times \mathbf{S}^1$.

Our goal is to average the summation above in the obstacle configuration $\{c\}$ under the Poisson distribution, and to identify a scaling on the obstacle radius r and the parameter n of the Poisson distribution leading to a nontrivial limit.

The parameter n has the following important physical interpretation. The expected number of obstacle centers to be found in any measurable subset Ω of the Euclidean plane $\mathbf{R}^2$ is

$$\sum_{p \geq 0} p\,\operatorname{Prob}(\{\{c\} \mid \#(\Omega \cap \{c\}) = p\}) = \sum_{p \geq 0} p e^{-n|\Omega|} \frac{(n|\Omega|)^p}{p!} = n|\Omega|,$$

so that

$$n = \#\text{ obstacles per unit surface in }\mathbf{R}^2.$$

The average of the first term in the summation defining $f(t, x, v, \{c\})$ is

$$f^{\mathrm{in}}(x - tv, v)\langle\mathbf{1}_{t < \tau_1}\rangle = f^{\mathrm{in}}(x - tv, v)e^{-n2rt}$$

(where $\langle\cdot\rangle$ denotes the mathematical expectation) since the condition $t < \tau_1$ means that the tube of width $2r$ and length t contains no obstacle center.

Henceforth, we seek a scaling limit corresponding to small obstacles, i.e., $r \to 0$ and a large number of obstacles per unit volume, i.e., $n \to \infty$.

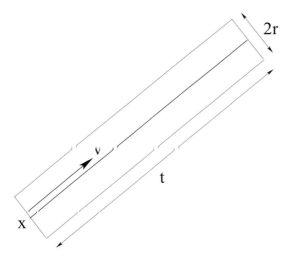

Figure 2.3: The tube corresponding with the first term in the series expansion giving the particle density

There are obviously many possible scalings that satisfy this requirement. Among all these scalings, the Boltzmann–Grad scaling in space dimension 2 is defined by the requirement that the average over obstacle configurations of the first term in the series expansion for the particle density f has a nontrivial limit.

BOLTZMANN–GRAD SCALING IN SPACE DIMENSION 2

In order for the average of the first term above to have a nontrivial limit, one must have

$$r \to 0^+ \text{ and } n \to +\infty \text{ in such a way that } 2nr \to \sigma > 0.$$

Under this assumption,

$$\langle f^{\text{in}}(x - tv, v) \, \mathbf{1}_{t < \tau_1} \rangle \longrightarrow f^{\text{in}}(x - tv, v)e^{-\sigma t}.$$

Gallavotti's idea is that this first term corresponds with the solution at time t of the equation

$$(\partial_t + v \cdot \nabla_x)f = -nrf \int_{\substack{|\omega|=1 \\ \omega \cdot v > 0}} \cos(v, \omega) \, d\omega = -2nrf,$$

$$f\big|_{t=0} = f^{\text{in}}$$

that involves only the loss part in the Lorentz collision integral, and that the (average over obstacle configuration of the) subsequent terms in the sum defining the particle density f should converge to the Duhamel formula for the Lorentz kinetic equation.

After these necessary preliminaries, we can state Gallavotti's theorem.

Theorem 2.2.1 (Gallavotti [19]). *Let f^{in} be a continuous, bounded probability density on $\mathbf{R}^2 \times \mathbf{S}^1$, and let*

$$f_r(t, x, v, \{c\}) = f^{\mathrm{in}}((X^r, V^r)(-t, x, v, \{c\})),$$

where $(t, x, v) \mapsto (X^r, V^r)(t, x, v, \{c\})$ is the billiard flow in the system of disks of radius r centered at the elements of $\{c\}$. Assuming that the obstacle centers are distributed under the Poisson law of parameter $n = \sigma/2r$ with $\sigma > 0$, the expected single particle density

$$\langle f_r(t, x, v, \cdot) \rangle \longrightarrow f(t, x, v) \ \ in \ L^1(\mathbf{R}^2 \times \mathbf{S}^1)$$

uniformly on compact t-sets, where f is the solution of the Lorentz kinetic equation

$$(\partial_t + v \cdot \nabla_x) f + \sigma f = \sigma \int_0^{2\pi} f(t, x, R[\beta] v) \sin \tfrac{\beta}{2} \, \tfrac{d\beta}{4} \, ,$$
$$f\big|_{t=0} = f^{\mathrm{in}},$$

where $R[\beta]$ denotes the rotation of an angle β.

End of the proof of Gallavotti's theorem. The general term in the summation giving $f(t, x, v, \{c\})$ is

$$f^{\mathrm{in}} \left(x - \sum_{k=1}^{j} \Delta \tau_k V^r(-\tau_k^-) - (t - \tau_j) V^r(-\tau_j^+), V^r(-\tau_j^+) \right) \mathbf{1}_{\tau_j < t < \tau_{j+1}},$$

and its average under the Poisson distribution on $\{c\}$ is

$$\int f^{\mathrm{in}} \left(x - \sum_{k=1}^{j} \Delta \tau_k V^r(-\tau_k^-) - (t - \tau_j) V^r(-\tau_j^+), V^r(-\tau_j^+) \right)$$
$$\times e^{-n|T(t; c_1, \ldots, c_j)|} \frac{n^j \, dc_1 \ldots dc_j}{j!},$$

where $T(t; c_1, \ldots, c_j)$ is the tube of width $2r$ around the particle trajectory colliding first with the obstacle centered at c_1 and whose j-th collision is with the obstacle centered at c_j.

As before, the surface of that tube is

$$|T(t; c_1, \ldots, c_j)| = 2rt + O(r^2).$$

In the j-th term, change variables by expressing the positions of the j encountered obstacles in terms of free flight times and deflection angles:

$$(c_1, \ldots, c_j) \longmapsto (\tau_1, \ldots, \tau_j; \beta_1, \ldots, \beta_j).$$

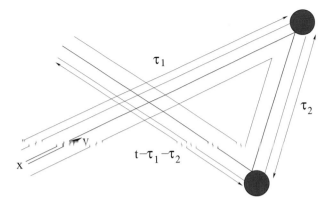

Figure 2.4: The tube $T(t, c_1, c_2)$ corresponding with the third term in the series expansion giving the particle density

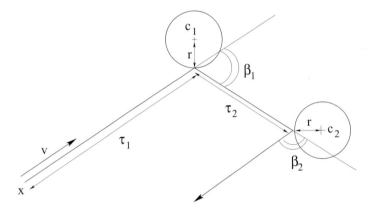

Figure 2.5: The substitution $(c_1, c_2) \mapsto (\tau_1, \tau_2, \beta_1, \beta_2)$

The volume element in the j-th integral is changed into

$$\frac{dc_1 \ldots dc_j}{j!} = r^j \sin\frac{\beta_1}{2} \cdots \sin\frac{\beta_j}{2} \frac{d\beta_1}{2} \cdots \frac{d\beta_j}{2} d\tau_1 \ldots d\tau_j.$$

The measure in the left-hand side is invariant by permutations of $c_1, \ldots, c_j$; on the right-hand side, we assume that

$$\tau_1 < \tau_2 < \cdots < \tau_j,$$

which explains why the $1/j!$ factor disappears in the right-hand side.

The substitution above is one-to-one only if the particle does not hit twice the same obstacle. Define therefore

$$A_r(T, x, v) = \{\{c\} \mid \text{there exists } 0 < t_1 < t_2 < T \text{ and } j \in \mathbf{N} \text{ such that}$$
$$\text{dist}(X^r(t_1, x, v, \{c\}), c_j) = \text{dist}(X^r(t_2, x, v, \{c\}), c_j) = r\}$$
$$= \bigcup_{j \geq 1} \{\{c\} \mid \text{dist}(X^r(t, x, v, \{c\}), c_j) = r \text{ for some } 0 < t_1 < t_2 < T\},$$

and set

$$f_r^M(t, x, v, \{c\}) = f_r(t, x, v, \{c\}) - f_r^R(t, x, v, \{c\}),$$
$$f_r^R(t, x, v, \{c\}) = f_r(t, x, v, \{c\})\, \mathbf{1}_{A_r(T, x, v)}(\{c\}),$$

respectively the Markovian part and the recollision part in f_r.

After averaging over the obstacle configuration $\{c\}$, the contribution of the j-th term in f_r^M is, to leading order in r:

$$(2nr)^j e^{-2nrt} \int_{0 < \tau_1 < \cdots < \tau_j < t} \int_{[0,2\pi]^j} \sin\frac{\beta_1}{2} \cdots \sin\frac{\beta_j}{2} \frac{d\beta_1}{4} \cdots \frac{d\beta_j}{4}\, d\tau_1 \ldots d\tau_j$$
$$\times f^{\text{in}}\left(x - \sum_{k=1}^{j}\Delta\tau_k R\left[\sum_{l=1}^{k-1}\beta_l\right]v - (t - \tau_j)R\left[\sum_{l=1}^{j}\beta_l\right]v, R\left[\sum_{l=1}^{j}\beta_l\right]v\right).$$

It is dominated by

$$\|f^{\text{in}}\|_{L^\infty} O(\sigma)^j e^{-O(\sigma)t}\frac{t^j}{j!}$$

which is the general term of a converging series.

Passing to the limit as $n \to +\infty$, $r \to 0$ so that $2rn \to \sigma$, one finds (by dominated convergence in the series) that

$$\langle f_r^M(t, x, v, \{c\})\rangle \longrightarrow e^{-\sigma t} f^{\text{in}}(x - tv, v)$$

$$+ \sigma e^{-\sigma t}\int_0^t \int_0^{2\pi} f^{\text{in}}(x - \tau_1 v - (t - \tau_1)R[\beta_1]v, R[\beta_1]v)\sin\frac{\beta_1}{2}\frac{d\beta_1}{4}\,d\tau_1$$

$$+ \sum_{j \geq 2}\sigma^j e^{-\sigma t}\int_{0 < \tau_j < \cdots < \tau_1 < t}\int_{[0,2\pi]^j}\sin\frac{\beta_1}{2}\cdots\sin\frac{\beta_j}{2}$$

$$\times f^{\text{in}}\left(x - \sum_{k=1}^{j}\Delta\tau_k R\left[\sum_{l=1}^{k-1}\beta_l\right]v - (t - \tau_j)R\left[\sum_{l=1}^{j}\beta_l\right]v, R\left[\sum_{l=1}^{j}\beta_l\right]v\right)$$

$$\times \frac{d\beta_1}{4}\cdots\frac{d\beta_j}{4}\,d\tau_1 \ldots d\tau_j,$$

which is the Duhamel series giving the solution of the Lorentz kinetic equation.

Hence, we have proved that

$$\langle f_r^M(t, x, v, \cdot)\rangle \to f(t, x, v) \text{ uniformly on bounded sets as } r \to 0^+,$$

where f is the solution of the Lorentz kinetic equation. One can check by a straight-forward computation that the Lorentz collision integral satisfies the property

$$\int_{\mathbf{S}^1} \mathcal{C}(\phi)(v)\, dv = 0 \text{ for each } \phi \in L^\infty(\mathbf{S}^1).$$

Integrating both sides of the Lorentz kinetic equation in the variables (t, x, v) over $[0, t] \times \mathbf{R}^2 \times \mathbf{S}^1$ shows that the solution f of that equation satisfies

$$\iint_{\mathbf{R}^2 \times \mathbf{S}^1} f(t, x, v) \, dx \, dv = \iint_{\mathbf{R}^2 \times \mathbf{S}^1} f^{\mathrm{in}}(x, v) \, dx \, dv$$

for each $t > 0$.

On the other hand, the billiard flow $(X, V)(t, \cdot, \cdot, \{c\})$ obviously leaves the uniform measure $dx\,dv$ on $\mathbf{R}^2 \times \mathbf{S}^1$ (i.e., the particle number) invariant, so that, for each $t > 0$ and each $r > 0$,

$$\iint_{\mathbf{R}^2 \times \mathbf{S}^1} f_r(t, x, v, \{c\}) \, dx \, dv = \iint_{\mathbf{R}^2 \times \mathbf{S}^1} f^{\mathrm{in}}(x, v) \, dx \, dv.$$

We therefore deduce from Fatou's lemma that

$$\langle f_r^R \rangle \to 0 \text{ in } L^1(\mathbf{R}^2 \times \mathbf{S}^1) \text{ uniformly on bounded } t\text{-sets, and}$$
$$\langle f_r^M \rangle \to f \text{ in } L^1(\mathbf{R}^2 \times \mathbf{S}^1) \text{ uniformly on bounded } t\text{-sets,}$$

which concludes our sketch of the proof of Gallavotti's theorem.   □

For a complete proof, we refer the interested reader to [19, 20].

Some remarks are in order before leaving Gallavotti's setting for the Lorentz gas with the Poisson distribution of obstacles.

Assuming no external force field as done everywhere in the present paper is not as inocuous as it may seem. For instance, in the case of Poisson distributed holes —i.e., purely absorbing obstacles, so that particles falling into the holes disappear from the system forever— the presence of an external force may introduce memory effects in the Boltzmann–Grad limit, as observed by L. Desvillettes and V. Ricci [15].

Another remark is about the method of proof itself. One has obtained the Lorentz kinetic equation *after* having obtained an explicit formula for the solution of that equation. In other words, the equation is deduced from the solution — which is a somewhat unusual situation in mathematics. However, the same is true of Lanford's derivation of the Boltzmann equation [29], as well as of the derivation of several other models in nonequilibrium statistical mechanics. For an interesting comment on this issue, see [13] on p. 75.

2.3   Santaló's formula for the geometric mean free path

From now on, we shall abandon the random case and concentrate our efforts on the periodic Lorentz gas.

Our first task is to define the Boltzmann–Grad scaling for periodic systems of spherical obstacles. In the Poisson case defined above, things were relatively easy: in space dimension 2, the Boltzmann–Grad scaling was defined by the prescription

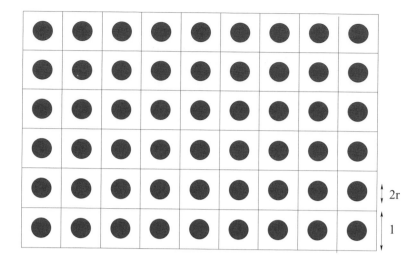

Figure 2.6: The periodic billiard table

that the number of obstacles per unit volume tends to infinity while the obstacle radius tends to 0 in such a way that

$$\text{\# obstacles per unit volume} \times \text{obstacle radius} \longrightarrow \sigma > 0.$$

The product above has an interesting geometric meaning even without assuming a Poisson distribution for the obstacle centers, which we shall briefly discuss before going further in our analysis of the periodic Lorentz gas.

Perhaps the most important scaling parameter in all kinetic models is the mean free path. This is by no means a trivial notion, as will be seen below. As suggested by the name itself, any notion of mean free path must involve first the notion of free path length, and then some appropriate probability measure under which the free path length is averaged.

For simplicity, the only periodic distribution of obstacles considered below is the set of balls of radius r centered at the vertices of a unit cubic lattice in the D-dimensional Euclidean space.

Correspondingly, for each $r \in (0, \frac{1}{2})$, we define the domain left free for particle motion, also called the "billiard table" as

$$Z_r = \{x \in \mathbf{R}^D \mid \text{dist}(x, \mathbf{Z}^D) > r\}.$$

Defining the free path length in the billiard table Z_r is easy: the free path length starting from $x \in Z_r$ in the direction $v \in \mathbf{S}^{D-1}$ is

$$\tau_r(x, v) = \min\{t > 0 \mid x + tv \in \partial Z_r\}.$$

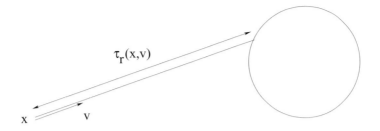

Figure 2.7: The free path length

Obviously, for each $v \in \mathbf{S}^{D-1}$ the free path length $\tau_r(\,\cdot\,,v)$ in the direction v can be extended continuously to

$$\{x \in \partial Z_r \mid v \cdot n_x \neq 0\},$$

where n_x denotes the unit normal vector to ∂Z_r at the point $x \in \partial Z_r$ pointing towards Z_r.

With this definition, the mean free path is the quantity defined as

$$\text{Mean Free Path} = \langle \tau_r \rangle,$$

where the notation $\langle \cdot \rangle$ designates the average under some appropriate probability measure on $\overline{Z_r} \times \mathbf{S}^{D-1}$.

A first ambiguity in the notion of mean free path comes from the fact that there are two fairly natural probability measures for the Lorentz gas.

The first one is the uniform probability measure on $(Z_r/\mathbf{Z}^D) \times \mathbf{S}^{D-1}$,

$$d\mu_r(x,v) = \frac{dx\,dv}{|Z_r/\mathbf{Z}^D|\,|\mathbf{S}^{D-1}|},$$

that is invariant under the billiard flow —the notation $|\mathbf{S}^{D-1}|$ designates the $(D-1)$-dimensional uniform measure of the unit sphere $\mathbf{S}^{D-1}$. This measure is obviously invariant under the billiard flow

$$(X_r, V_r)(t, \cdot\,, \cdot) : Z_r \times \mathbf{S}^{D-1} \longrightarrow Z_r \times \mathbf{S}^{D-1}$$

defined by

$$\begin{cases} \dot{X}_r = V_r \\ \dot{V}_r = 0 \end{cases} \quad \text{whenever } X(t) \notin \partial Z_r$$

while

$$\begin{cases} X_r(t^+) = X_r(t^-) =: X_r(t) \text{ if } X(t^{\pm}) \in \partial Z_r, \\ V_r(t^+) = \mathcal{R}[n_{X_r(t)}]V_r(t^-), \end{cases}$$

with $\mathcal{R}[n]v = v - 2(v \cdot n)n$ denoting the reflection with respect to the hyperplane $(\mathbf{R}n)^{\perp}$.

The second such probability measure is the invariant measure of the billiard map

$$d\nu_r(x, v) = \frac{(v \cdot n_x)_+ \, dS(x) \, dv}{(v \cdot n_x)_+ \, dx \, dv\text{-meas}(\Gamma_+^r/\mathbf{Z}^D)}$$

where n_x is the unit inward normal at $x \in \partial Z_r$, while $dS(x)$ is the $(D-1)$-dimensional surface element on ∂Z_r, and

$$\Gamma_+^r = \{(x, v) \in \partial Z_r \times \mathbf{S}^{D-1} \mid v \cdot n_x > 0\}.$$

The billiard map $\mathcal{B}_r$ is the map

$$\Gamma_+^r \ni (x, v) \longmapsto \mathcal{B}_r(x, v) = (X_r, V_r)(\tau_r(x, v); x, v) \in \Gamma_+^r,$$

which obviously passes to the quotient modulo $\mathbf{Z}^D$-translations:

$$\mathcal{B}_r : \; \Gamma_+^r/\mathbf{Z}^D \longrightarrow \Gamma_+^r/\mathbf{Z}^D.$$

In other words, given the position x and the velocity v of a particle immediately after its first collision with an obstacle, the sequence $(\mathcal{B}_r^n(x, v))_{n \geq 0}$ is the sequence of all collision points and post-collision velocities on that particle's trajectory.

With the material above, we can define a first, very natural notion of mean free path, by setting

$$\text{Mean Free Path} = \lim_{N \to +\infty} \frac{1}{N} \sum_{k=0}^{N-1} \tau_r(\mathcal{B}_r^k(x, v)).$$

Notice that, for ν_r-a.e. $(x, v) \in \Gamma_r^+/\mathbf{Z}^D$, the right-hand side of the equality above is well-defined by the Birkhoff ergodic theorem. If the billiard map $\mathcal{B}_r$ is ergodic for the measure ν_r, one has

$$\lim_{N \to +\infty} \frac{1}{N} \sum_{k=0}^{N-1} \tau_r(\mathcal{B}_r^k(x, v)) = \int_{\Gamma_+^r/\mathbf{Z}^D} \tau_r \, d\nu_r,$$

for ν_r-a.e. $(x, v) \in \Gamma_+^r/\mathbf{Z}^D$.

Now, a very general formula for computing the right-hand side of the above equality was found by the great Spanish mathematician L. A. Santaló in 1942. In fact, Santaló's argument applies to situations that are considerably more general, involving for instance curved trajectories instead of straight line segments, or obstacle distributions other than periodic. The reader interested in these questions is referred to Santaló's original article [38].

Figure 2.8: Luis Antonio Santaló Sors (1911–2001)

SANTALÓ'S FORMULA FOR THE GEOMETRIC MEAN FREE PATH

One finds that

$$\ell_r = \int_{\Gamma_+^r/\mathbf{Z}^D} \tau_r(x,v)\, d\nu_r(x,v) = \frac{1 - |\mathbf{B}^D|r^D}{|\mathbf{B}^{D-1}|r^{D-1}}$$

where $\mathbf{B}^D$ is the unit ball of $\mathbf{R}^D$ and $|\mathbf{B}^D|$ its D-dimensional Lebesgue measure.

In fact, one has the following slightly more general

Lemma 2.3.1 (H. S. Dumas, L. Dumas, F. Golse [17]). *For $f \in C^1(\mathbf{R}_+)$ such that $f(0) = 0$, one has*

$$\iint_{\Gamma_+^r/\mathbf{Z}^D} f(\tau_r(x,v))v \cdot n_x\, dS(x)\, dv = \iint_{(Z_r/\mathbf{Z}^D) \times \mathbf{S}^{D-1}} f'(\tau_r(x,v))\, dx\, dv.$$

Santaló's formula is obtained by setting $f(z) = z$ in the identity above, and expressing both integrals in terms of the normalized measures ν_r and μ_r.

Proof. For each $(x,v) \in Z_r \times \mathbf{S}^{D-1}$ one has

$$\tau_r(x + tv, v) = \tau_r(x,v) - t,$$

so that

$$\frac{d}{dt}\tau_r(x+tv,v)=-1.$$

Hence $\tau_r(x,v)$ solves the transport equation

$$\begin{cases} v\cdot\nabla_x\tau_r(x,v)=-1, & x\in Z_r, \quad v\in \mathbf{S}^{D-1}, \\ \tau_r(x,v)=0, & x\in\partial Z_r, \quad v\cdot n_x<0. \end{cases}$$

Since $f\in C^1(\mathbf{R}_+)$ and $f(0)=0$, one has

$$\begin{cases} v\cdot\nabla_x f(\tau_r(x,v))=-f'(\tau_r(x,v)), & x\in Z_r, \quad v\in\mathbf{S}^{D-1}, \\ f(\tau_r(x,v))=0, & x\in\partial Z_r, \quad v\cdot n_x<0. \end{cases}$$

Integrating both sides of the equality above, and applying Green's formula shows that

$$-\iint_{(Z_r/\mathbf{Z}^D)\times\mathbf{S}^{D-1}}f'(\tau_r(x,v))\,dx\,dv$$

$$=\iint_{(Z_r/\mathbf{Z}^D)\times\mathbf{S}^{D-1}}v\cdot\nabla_x(f(\tau_r(x,v)))\,dx\,dv$$

$$=-\iint_{(\partial Z_r/\mathbf{Z}^D)\times\mathbf{S}^{D-1}}f(\tau_r(x,v))v\cdot n_x\,dS(x)\,dv.$$

Beware the unusual sign in the right-hand side of the second equality above, coming from the orientation of the unit normal n_x, which is pointing towards Z_r.   □

With the help of Santaló's formula, we define the Boltzmann–Grad limit for the Lorentz gas with periodic as well as random distribution of obstacles as follows:

BOLTZMANN–GRAD SCALING

The Boltzmann–Grad scaling for the periodic Lorentz gas in space dimension D corresponds with the following choice of parameters:

$$\text{distance between neighboring lattice points}=\varepsilon\ll 1,$$

$$\text{obstacle radius}=r\ll 1,$$

$$\text{mean free path}=\ell_r\to\frac{1}{\sigma}>0.$$

Santaló's formula indicates that one should have

$$r\sim c\varepsilon^{\frac{D}{D-1}}\text{ with }c=\left(\frac{\sigma}{|\mathbf{B}^{D-1}|}\right)^{-\frac{1}{D-1}}\text{ as }\varepsilon\to 0^+.$$

Therefore, given an initial particle density $f^{\text{in}}\in C_c(\mathbf{R}^D\times\mathbf{S}^{D-1})$, we define f_r to be

$$f_r(t,x,v)=f^{\text{in}}\left(r^{D-1}X_r\left(-\frac{t}{r^{D-1}};\frac{x}{r^{D-1}},v\right),V_r\left(-\frac{t}{r^{D-1}};\frac{x}{r^{D-1}},v\right)\right)$$

where (X_r, V_r) is the billiard flow in Z_r with specular reflection on ∂Z_r.

Notice that this formula defines f_r for $x \in Z_r$ only, as the particle density should remain 0 for all time in the spatial domain occupied by the obstacles. As explained in the previous section, this is a set whose measure vanishes in the Boltzmann–Grad limit, and we shall always implicitly extend the function f_r defined above by 0 for $x \notin Z_r$.

Since $f^{\rm in}$ is a bounded function on $Z_r \times \mathbf{S}^{D-1}$, the family f_r defined above is a bounded family of $L^\infty(\mathbf{R}^D \times \mathbf{S}^{D-1})$. By the Banach–Alaoglu theorem, this family is therefore relatively compact for the weak-* topology of $L^\infty(\mathbf{R}_+ \times \mathbf{R}^D \times \mathbf{S}^{D-1})$.

Problem: Find an equation governing the L^∞ weak-* limit points of the scaled number density f_r as $r \to 0^+$.

In the sequel, we shall describe the answer to this question in the 2-dimensional case $(D = 2)$.

2.4 Estimates for the distribution of free-path lengths

In the proof of Gallavotti's theorem for the case of a Poisson distribution of obstacles in space dimension $D = 2$, the probability that a strip of width $2r$ and length t does not meet any obstacle is e^{-2nrt}, where n is the parameter of the Poisson distribution —i.e., the average number of obstacles per unit surface.

This accounts for the loss term

$$f^{\rm in}(x - tv, v)e^{-\sigma t}$$

in the Duhamel series for the solution of the Lorentz kinetic equation, or of the term $-\sigma f$ on the right-hand side of that equation written in the form

$$(\partial_t + v \cdot \nabla_x)f = -\sigma f + \sigma \int_0^{2\pi} f(t, x, R(\beta)v) \sin \tfrac{\beta}{2} \tfrac{d\beta}{4}.$$

Things are fundamentally different in the periodic case. To begin with, there are infinite strips included in the billiard table Z_r which *never* meet any obstacle.

The contribution of the 1-particle density leading to the loss term in the Lorentz kinetic equation is, in the notation of the proof of Gallavotti's theorem,

$$f^{\rm in}(x - tv, v)\, \mathbf{1}_{t < \tau_1(x, v, \{c\})}.$$

The analogous term in the periodic case is

$$f^{\rm in}(x - tv, v)\, \mathbf{1}_{t < r^{D-1}\tau_r(x/r^{D-1}, -v)}$$

where $\tau_r(x, v)$ is the free-path length in the periodic billiard table Z_r starting from $x \in Z_r$ in the direction $v \in \mathbf{S}^1$.

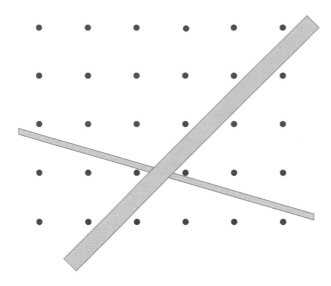

Figure 2.9: Open strips in the periodic billiard table that never meet any obstacle

Passing to the L^∞ weak-* limit as $r \to 0$ reduces to finding

$$\lim_{r \to 0} \mathbf{1}_{t < r^{D-1} \tau_r(x/r^{D-1}, -v)} \text{ in } w^* - L^\infty(\mathbf{R}^2 \times \mathbf{S}^1)$$

—possibly after extracting a subsequence $r_n \downarrow 0$. As we shall see below, this involves the distribution of τ_r under the probability measure μ_r introduced in the discussion of Santaló's formula —i.e., assuming the initial position x and direction v to be independent and uniformly distributed on $(\mathbf{R}^D/\mathbf{Z}^D) \times \mathbf{S}^{D-1}$.

We define the (scaled) distribution under μ_r of free path lengths τ_r to be

$$\Phi_r(t) = \mu_r(\{(x, v) \in (Z_r/\mathbf{Z}^D) \times \mathbf{S}^{D-1} \mid \tau_r(x, v) > t/r^{D-1}\}).$$

Notice the scaling $t \mapsto t/r^{D-1}$ in this definition. In space dimension D, Santaló's formula shows that

$$\iint_{\Gamma_r^+/\mathbf{Z}^D} \tau_r(x, v) \, d\nu_r(x, v) \sim \frac{1}{|\mathbf{B}^{D-1}|} r^{1-D},$$

and this suggests that the free path length τ_r is a quantity of the order of $1/r^{D-1}$. (In fact, this argument is not entirely convincing, as we shall see below.)

In any case, with this definition of the distribution of free path lengths under μ_r, one arrives at the following estimate.

Theorem 2.4.1 (Bourgain–Golse–Wennberg [6, 25]). *In space dimension $D \geq 2$, there exist $0 < C_D < C_D'$ such that*

$$\frac{C_D}{t} \leq \Phi_r(t) \leq \frac{C_D'}{t} \quad \text{whenever } t > 1 \text{ and } 0 < r < \tfrac{1}{2}.$$

The lower bound and the upper bound in this theorem are obtained by very different means.

The upper bound follows from a Fourier series argument which is reminiscent of Siegel's proof of the classical Minkowski convex body theorem (see [39, 36]).

The lower bound, on the other hand, is obtained by working in physical space. Specifically, one uses a channel technique, introduced independently by P. Bleher [2] for the diffusive scaling.

This lower bound alone has an important consequence:

Corollary 2.4.2. *For each $r > 0$, the average of the free path length (mean free path) under the probability measure μ_r is infinite:*

$$\int_{(Z_r/\mathbf{Z}^D)\times\mathbf{S}^{D-1}} \tau_r(x,v)\, d\mu_r(x,v) = +\infty.$$

Proof. Indeed, since Φ_r is the distribution of τ_r under μ_r, one has

$$\int_{(Z_r/\mathbf{Z}^D)\times\mathbf{S}^{D-1}} \tau_r(x,v)\, d\mu_r(x,v) = \int_0^\infty \Phi_r(t)\, dt \geq \int_1^\infty \frac{C_D}{t}\, dt = +\infty. \qquad \square$$

Recall that the average of the free path length under the "other" natural probability measure ν_r is precisely Santaló's formula for the mean free path:

$$\ell_r = \iint_{\Gamma_r^+/\mathbf{Z}^D} \tau_r(x,v)\, d\nu_r(x,v) = \frac{1 - |\mathbf{B}^D| r^D}{|\mathbf{B}^{D-1}| r^{D-1}}.$$

One might wonder why averaging the free path length τ_r under the measures ν_r and μ_r actually gives two so different results.

First observe that Santaló's formula gives the mean free path under the probability measure ν_r concentrated on the surface of the obstacles, and is therefore irrelevant for particles that have not yet encountered an obstacle.

Besides, by using the lemma that implies Santaló's formula with $f(z) = \frac{1}{2}z^2$, one has

$$\iint_{(Z_r/\mathbf{Z}^D)\times\mathbf{S}^{D-1}} \tau_r(x,v)\, d\mu_r(x,v) = \frac{1}{\ell_r} \iint_{\Gamma_r^+/\mathbf{Z}^D} \tfrac{1}{2}\tau_r(x,v)^2\, d\nu_r(x,v).$$

Whenever the components $v_1,\ldots,v_D$ are independent over $\mathbf{Q}$, the linear flow in the direction v is topologically transitive and ergodic on the D-torus, so that $\tau_r(x,v) < +\infty$ for each $r > 0$ and $x \in \mathbf{R}^D$. On the other hand, $\tau_r(x,v) = +\infty$ for some $x \in Z_r$ (the periodic billiard table) whenever v belongs to some specific class of unit vectors whose components are rationally dependent, a class that becomes dense in $\mathbf{S}^{D-1}$ as $r \to 0^+$. Thus, τ_r is strongly oscillating (finite for irrational directions, possibly infinite for a class of rational directions that becomes dense as $r \to 0^+$), and this explains why τ_r does not have a second moment under ν_r.

Proof of the Bourgain–Golse–Wennberg lower bound. We shall restrict our attention to the case of space dimension $D = 2$.

As mentioned above, there are *infinite nonempty open strips* included in Z_r —i.e., never meeting any obstacle. Call *a channel* any such nonempty open strip of maximum width, and let C_r be the set of all channels included in Z_r.

If $S \in C_r$ and $x \in S$, define $\tau_S(x, v)$ the exit time from the channel starting from x in the direction v, defined as

$$\tau_S(x, v) = \inf\{t > 0 \mid x + tv \in \partial S\}, \quad (x, v) \in S \times \mathbf{S}^1.$$

Obviously, any particle starting from x in the channel S in the direction v must exit S before it hits an obstacle (since no obstacle intersects S). Therefore

$$\tau_r(x, v) \geq \sup\{\tau_S(x, v) \mid S \in C_r \text{ such that } x \in S\},$$

so that

$$\Phi_r(t) \geq \mu_r \left(\bigcup_{S \in C_r} \{(x, v) \in (S/\mathbf{Z}^2) \times \mathbf{S}^1 \mid \tau_S(x, v) > t/r\} \right).$$

This observation suggests that one should carefully study the set of channels C_r.

Step 1: Description of C_r. Given $\omega \in \mathbf{S}^1$, we define

$$C_r(\omega) = \{\text{channels of direction } \omega \text{ in } C_r\}.$$

We begin with a lemma which describes the structure of $C_r(\omega)$.

Lemma 2.4.3. *Let $r \in [0, \frac{1}{2})$ and $\omega \in \mathbf{S}^1$. Then:*

1) *if $S \in C_r(\omega)$, then*

$$C_r(\omega) = \{S + k \mid k \in \mathbf{Z}^2\};$$

2) *if $C_r(\omega) \neq \emptyset$, then*

$$\omega = \frac{(p, q)}{\sqrt{p^2 + q^2}}$$

with

$$(p, q) \in \mathbf{Z}^2 \setminus \{(0, 0)\} \text{ such that } \gcd(p, q) = 1 \text{ and } \sqrt{p^2 + q^2} < \frac{1}{2r}.$$

We henceforth denote by A_r the set of all such $\omega \in \mathbf{S}^1$. Then

3) *for $\omega \in A_r$, the elements of $C_r(\omega)$ are open strips of width*

$$w(\omega, r) = \frac{1}{\sqrt{p^2 + q^2}} - 2r.$$

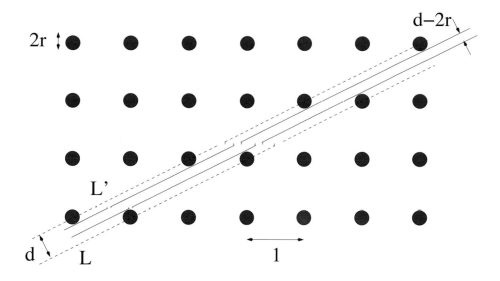

Figure 2.10: A channel of direction $\omega = \frac{1}{\sqrt{5}}(2,1)$; minimal distance d between lines L and L' of direction ω through lattice points

Proof of the lemma. Statement 1) is obvious. As for statement 2), if L is an infinite line of direction $\omega \in \mathbf{S}^1$ such that ω_2/ω_1 is irrational, then $L/\mathbf{Z}^2$ is an orbit of a linear flow on $\mathbf{T}^2$ with irrational slope ω_2/ω_1. Therefore $L/\mathbf{Z}^2$ is dense in $\mathbf{T}^2$ so that L cannot be included in Z_r.

Assume that

$$\omega = \frac{(p,q)}{\sqrt{p^2 + q^2}} \text{ with } (p,q) \in \mathbf{Z}^2 \setminus \{(0,0)\} \text{ coprime,}$$

and let L, L' be two infinite lines with direction ω, with equations

$$qx - py = a \text{ and } qx - py = a' \text{ respectively.}$$

Obviously

$$\mathrm{dist}(L, L') = \frac{|a - a'|}{\sqrt{p^2 + q^2}}.$$

If $L \cup L'$ is the boundary of a channel of direction

$$\omega = \frac{(p,q)}{\sqrt{p^2 + q^2}} \in \mathcal{A}_0$$

included in $\mathbf{R}^2 \setminus \mathbf{Z}^2$ —i.e., of an element of $\mathcal{C}_0(\omega)$, then L and L' intersect $\mathbf{Z}^2$ so that

$$a, a' \in p\mathbf{Z} + q\mathbf{Z} = \mathbf{Z}$$

—the equality above following from the assumption that p and q are coprime. Since $\mathrm{dist}(L, L') > 0$ is minimal, then $|a - a'| = 1$, so that

$$\mathrm{dist}(L, L') = \frac{1}{\sqrt{p^2 + q^2}}.$$

Likewise, if $L \cup L' = \partial S$ with $S \in \mathcal{C}_r$, then L and L' are parallel infinite lines tangent to ∂Z_r, and the minimal distance between any such distinct lines is

$$\mathrm{dist}(L, L') = \frac{1}{\sqrt{p^2 + q^2}} - 2r.$$

This entails 2) and 3).   □

Step 2: The exit time from a channel. Let $\omega = \frac{(p,q)}{\sqrt{p^2+q^2}} \in \mathcal{A}_r$ and let $S \in \mathcal{C}_r(\omega)$. Cut S into three parallel strips of equal width and call $\hat{S}$ the middle one. For each $t > 1$ define

$$\theta \equiv \theta(\omega, r, t) = \arcsin\left(\frac{rw(\omega, r)}{3t}\right).$$

Lemma 2.4.4. *If* $x \in \hat{S}$ *and* $v \in (R[-\theta]\omega, R[\theta]\omega)$, *where* $R[\theta]$ *designates the rotation of an angle* θ, *then*

$$\tau_S(x, v) \geq t/r.$$

Moreover

$$\mu_r((\hat{S}/\mathbf{Z}^2) \times (R[-\theta]\omega, R[\theta]\omega)) = \tfrac{2}{3}w(\omega, r)\theta(\omega, r, t).$$

The proof of this lemma is perhaps best explained by considering Figure 2.11.

Step 3: Putting all channels together. Recall that we need to estimate

$$\mu_r\left(\bigcup_{S \in \mathcal{C}_r} \{(x, v) \in (S/\mathbf{Z}^2) \times \mathbf{S}^1 \mid \tau_S(x, v) > t/r\}\right).$$

Pick

$$\mathcal{A}_r \ni \omega = \frac{(p, q)}{\sqrt{p^2 + q^2}} \neq \frac{(p', q')}{\sqrt{p'^2 + q'^2}} = \omega' \in \mathcal{A}_r.$$

Observe that

$$|\sin(\widehat{\omega, \omega'})| = \frac{|pq' - p'q|}{\sqrt{p^2 + q^2}\sqrt{p'^2 + q'^2}} \geq \frac{1}{\sqrt{p^2 + q^2}\sqrt{p'^2 + q'^2}}$$

$$\geq \max\left(\frac{2r}{\sqrt{p^2 + q^2}}, \frac{2r}{\sqrt{p'^2 + q'^2}}\right) \geq \sin\theta(\omega, r, t) + \sin\theta(\omega', r, t)$$

$$\geq \sin(\theta(\omega, r, t) + \theta(\omega', r, t))$$

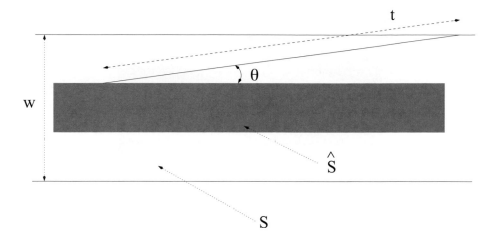

Figure 2.11: Exit time from the middle third $\hat{S}$ of an infinite strip S of width w

whenever $t > 1$.

Then, whenever $S \in \mathcal{C}_r(\omega)$ and $S' \in \mathcal{C}_r(\omega')$,

$$(\hat{S} \times (R[-\theta]\omega, R[\theta]\omega))) \cap (\hat{S}' \times (R[-\theta']\omega', R[\theta']\omega'))) = \varnothing$$

with $\theta = \theta(\omega, r, t)$, $\theta' = \theta'(\omega', r, t)$ and $R[\theta] = $ rotation of an angle θ.

Moreover, if $\omega = \frac{(p,q)}{\sqrt{p^2+q^2}} \in \mathcal{A}_r$ then

$$|\hat{S}/\mathbf{Z}^2| = \tfrac{1}{3} w(\omega, r)\sqrt{p^2 + q^2},$$

while

$$\#\{S/\mathbf{Z}^2 \mid S \in \mathcal{C}_r(\omega)\} = 1.$$

Conclusion: Therefore, whenever $t > 1$,

$$\bigcup_{S \in \mathcal{C}_r} (\hat{S}/\mathbf{Z}^2) \times (R[-\theta]\omega, R[\theta]\omega)$$

$$\subset \bigcup_{S \in \mathcal{C}_r} \{(x, v) \in (S/\mathbf{Z}^2) \times \mathbf{S}^1 \mid \tau_S(x, v) > t/r\},$$

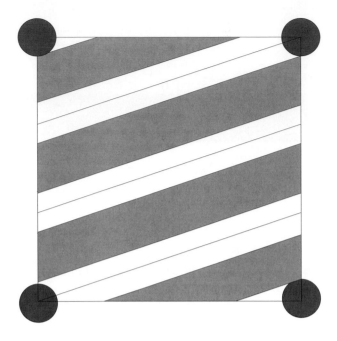

Figure 2.12: A channel modulo $\mathbf{Z}^2$

and the left-hand side is a disjoint union. Hence,

$$\mu_r\left(\bigcup_{S\in\mathcal{C}_r}\{(x,v)\in(S/\mathbf{Z}^2)\times\mathbf{S}^1\mid\tau_S(x,v)>t/r\}\right)$$

$$\geq\sum_{\omega\in\mathcal{A}_r}\mu_r((\hat{S}/\mathbf{Z}^2)\times(R[-\theta]\omega,R[\theta]\omega))$$

$$=\sum_{\substack{\gcd(p,q)=1\\p^2+q^2<1/4r^2}}\tfrac{1}{3}w(\omega,r)\sqrt{p^2+q^2}\cdot2\theta(\omega,r,t)$$

$$=\sum_{\substack{\gcd(p,q)=1\\p^2+q^2<1/4r^2}}\tfrac{2}{3}\sqrt{p^2+q^2}\,w(\omega,r)\arcsin\left(\frac{rw(\omega,r)}{3t}\right)$$

$$\geq\sum_{\substack{\gcd(p,q)=1\\p^2+q^2<1/4r^2}}\tfrac{2}{3}\sqrt{p^2+q^2}\,\frac{rw(\omega,r)^2}{3t}.$$

Now $\sqrt{p^2+q^2}<1/4r$ if and only if $w(\omega,r)=\frac{1}{\sqrt{p^2+q^2}}-2r>\frac{1}{2\sqrt{p^2+q^2}}$, so that, eventually,

$$\Phi_r(t) \geq \sum_{\substack{\gcd(p,q)=1 \\ p^2+q^2<1/16r^2}} \frac{2}{3}\sqrt{p^2+q^2}\,\frac{rw(\omega,r)^2}{3t} \geq \frac{r^2}{18t} \sum_{\substack{\gcd(p,q)=1 \\ p^2+q^2<1/16r^2}} \left[\frac{1}{r\sqrt{p^2+q^2}}\right].$$

This gives the desired conclusion, since

$$\sum_{\substack{\gcd(p,q)=1 \\ p^2+q^2<1/16r^2}} \left[\frac{1}{4r\sqrt{p^2+q^2}}\right] = \sum_{p^2+q^2<1/16r^2} 1 \sim \frac{\pi}{16r^2}.$$

The equality above is proved as follows: the term

$$\left[\frac{1}{4r\sqrt{p^2+q^2}}\right]$$

is the number of integer points on the segment of length $1/4r$ in the direction (p,q) with $(p,q) \in \mathbf{Z}^2$ such that $\gcd(p,q) = 1$.

The Bourgain–Golse–Wennberg theorem raises the question of whether $\Phi_r(t) \simeq C/t$ in some sense as $r \to 0^+$ and $t \to +\infty$. Given the very different nature of the arguments used to establish the upper and the lower bounds in that theorem, this is a highly nontrivial problem, whose answer seems to be known only in space dimension $D = 2$ so far. We shall return to this question later, and see that the 2-dimensional situation is amenable to a class of very specific techniques based on continued fractions, that can be used to encode particle trajectories of the periodic Lorentz gas.

A first answer to this question, in space dimension $D = 2$, is given by the following

Theorem 2.4.5 (Caglioti–Golse [9]). *Assume $D = 2$ and define, for each $v \in \mathbf{S}^1$,*

$$\phi_r(t|v) = \mu_r(\{x \in Z_r/\mathbf{Z}^2 \mid \tau_r(x,v) \geq t/r\}), \quad t \geq 0.$$

Then there exists $\Phi\colon \mathbf{R}_+ \to \mathbf{R}_+$ such that

$$\frac{1}{|\ln\varepsilon|} \int_\varepsilon^{1/4} \phi_r(t,v)\frac{dr}{r} \longrightarrow \Phi(t) \ a.e. \ in \ v \in \mathbf{S}^1$$

in the limit as $\varepsilon \to 0^+$. Moreover,

$$\Phi(t) \sim \frac{1}{\pi^2 t} \ as \ t \to +\infty.$$

Shortly after [9] appeared, F. Boca and A. Zaharescu improved our method and managed to compute $\Phi(t)$ explicitly for each $t \geq 0$. One should keep in mind that their formula had been conjectured earlier by P. Dahlqvist [14], on the basis of a formal computation.

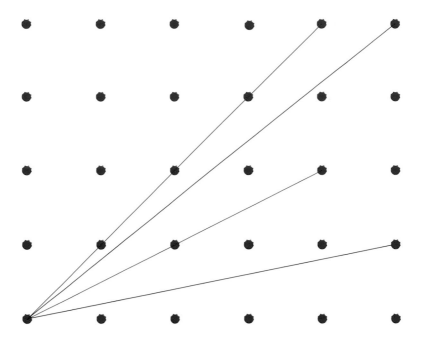

Figure 2.13: Black lines issued from the origin terminate at integer points with coprime coordinates; red lines terminate at integer points whose coordinates are not coprime

Theorem 2.4.6 (Boca–Zaharescu [3]). *For each $t > 0$,*

$$\Phi_r(t) \longrightarrow \Phi(t) = \frac{6}{\pi^2} \int_t^\infty (s - t)g(s)\, ds$$

in the limit as $r \to 0^+$, where

$$g(s) = \begin{cases} 1 & \text{if } s \in [0,1], \\ \frac{1}{s} + 2\left(1 - \frac{1}{s}\right)^2 \ln\left(1 - \frac{1}{s}\right) - \frac{1}{2}\left|1 - \frac{2}{s}\right|^2 \ln\left|1 - \frac{2}{s}\right| & \text{if } s \in (1, \infty). \end{cases}$$

In the sequel, we shall return to the continued and Farey fractions techniques used in the proofs of these two results, and generalize them.

2.5   A negative result for the Boltzmann–Grad limit of the periodic Lorentz gas

The material at our disposal so far provides us with a first answer —albeit a negative one— to the problem of determining the Boltzmann–Grad limit of the periodic Lorentz gas.

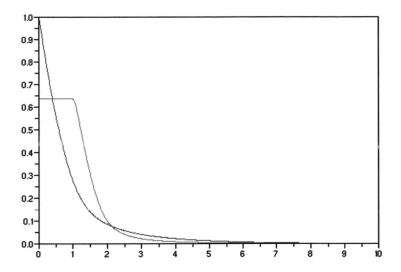

Figure 2.14: Graph of $\Phi(t)$ (blue curve) and $\Phi''(t)$ (green curve)

For simplicity, we consider the case of a Lorentz gas enclosed in a periodic box $\mathbf{T}^D = \mathbf{R}^D/\mathbf{Z}^D$ of unit side. The distance between neighboring obstacles is supposed to be ε^{D-1} with $0 < \varepsilon = 1/n$, for $n \in \mathbf{N}$ and $n > 2$ so that $\varepsilon < 1/2$, while the obstacle radius is $\varepsilon^D < \frac{1}{2}\varepsilon^{D-1}$ —so that obstacles never overlap. Define

$$Y_\varepsilon = \{x \in \mathbf{T}^D \mid \mathrm{dist}(x, \varepsilon^{D-1}\mathbf{Z}^D) > \varepsilon^D\} = \varepsilon^{D-1}(Z_\varepsilon/\mathbf{Z}^D).$$

For each $f^{\mathrm{in}} \in C(\mathbf{T}^D \times \mathbf{S}^{D-1})$, let f_ε be the solution of

$$\partial_t f_\varepsilon + v \cdot \nabla_x f_\varepsilon = 0, \quad (x,v) \in Y_\varepsilon \times \mathbf{S}^{D-1},$$
$$f_\varepsilon(t,x,v) = f_\varepsilon(t,x,\mathcal{R}[n_x]v), \quad (x,v) \in \partial Y_\varepsilon \times \mathbf{S}^{D-1},$$
$$f_\varepsilon\big|_{t=0} = f^{\mathrm{in}},$$

where n_x is unit normal vector to ∂Y_ε at the point x, pointing towards the interior of Y_ε.

By the method of characteristics,

$$f_\varepsilon(t,x,v) = f^{\mathrm{in}}\left(\varepsilon^{D-1} X_\varepsilon\left(-\frac{t}{\varepsilon^{D-1}}; \frac{x}{\varepsilon^{D-1}}, v\right); V_\varepsilon\left(-\frac{t}{\varepsilon^{D-1}}; \frac{x}{\varepsilon^{D-1}}, v\right)\right),$$

where $(X_\varepsilon, V_\varepsilon)$ is the billiard flow in Z_ε.

The main result in this section is the following.

Theorem 2.5.1 (Golse [21, 24]). *There exist initial data $f^{in} \equiv f^{in}(x) \in C(\mathbf{T}^D)$ such that no subsequence of f_ε converges in $L^\infty(\mathbf{R}_+ \times \mathbf{T}^D \times \mathbf{S}^{D-1})$ weak-$*$ to the solution f of a linear Boltzmann equation of the form*

$$(\partial_t + v \cdot \nabla_x)f(t,x,v) = \sigma \int_{\mathbf{S}^{D-1}} p(v,v')(f(t,x,v') - f(t,x,v))\,dv',$$

$$f\big|_{t=0} = f^{in},$$

where $\sigma > 0$ and $0 \le p \in L^2(\mathbf{S}^{D-1} \times \mathbf{S}^{D-1})$ satisfies

$$\int_{\mathbf{S}^{D-1}} p(v,v')\,dv' = \int_{\mathbf{S}^{D-1}} p(v',v)\,dv' = 1 \text{ a.e. in } v \in \mathbf{S}^{D-1}.$$

This theorem has the following important —and perhaps surprising— consequence: *the Lorentz kinetic equation cannot govern the Boltzmann–Grad limit of the particle density in the case of a periodic distribution of obstacles.*

Proof. The proof of the negative result above involves two different arguments:

a) the existence of a spectral gap for any linear Boltzmann equation, and

b) the lower bound for the distribution of free path lengths in the Bourgain–Golse–Wennberg theorem.

Step 1: Spectral gap for the linear Boltzmann equation

With $\sigma > 0$ and p as above, consider the unbounded operator A on $L^2(\mathbf{T}^D \times \mathbf{S}^{D-1})$ defined by

$$(A\phi)(x,v) = -v \cdot \nabla_x \phi(x,v) - \sigma\phi(x,v) + \sigma \int_{\mathbf{S}^{D-1}} p(v,v')\phi(x,v')\,dv',$$

with domain

$$D(A) = \{\phi \in L^2(\mathbf{T}^D \times \mathbf{S}^{D-1}) \mid v \cdot \nabla_x \phi \in L^2(\mathbf{T}^D \times \mathbf{S}^{D-1})\}.$$

Then:

Theorem 2.5.2 (Ukai–Point–Ghidouche [45]). *There exist positive constants C and γ such that*

$$\|e^{tA}\phi - \langle\phi\rangle\|_{L^2(\mathbf{T}^D \times \mathbf{S}^{D-1})} \le Ce^{-\gamma t}\|\phi\|_{L^2(\mathbf{T}^D \times \mathbf{S}^{D-1})}, \quad t \ge 0,$$

for each $\phi \in L^2(\mathbf{T}^D \times \mathbf{S}^{D-1})$, where

$$\langle\phi\rangle = \frac{1}{|\mathbf{S}^{D-1}|} \iint_{\mathbf{T}^D \times \mathbf{S}^{D-1}} \phi(x,v)\,dx\,dv.$$

Taking this theorem for granted, we proceed to the next step in the proof, leading to an explicit lower bound for the particle density.

Step 2: Comparison with the case of absorbing obstacles

Assume that $f^{\mathrm{in}} \equiv f^{\mathrm{in}}(x) \geq 0$ on $\mathbf{T}^D$. Then

$$f_\varepsilon(t,x,v) \geq g_\varepsilon(t,x,v) = f^{\mathrm{in}}(x-tv)\,\mathbf{1}_{Y_\varepsilon}(x)\,\mathbf{1}_{\varepsilon^{D-1}\tau_\varepsilon(x/\varepsilon^{D-1},v)>t}.$$

Indeed, g is the density of particles with the *same* initial data as f, but assuming that each particle *disappears* when colliding with an obstacle instead of being reflected.

Then

$$\mathbf{1}_{Y_\varepsilon}(x) \to 1 \text{ a.e. on } \mathbf{T}^D \text{ and } |\mathbf{1}_{Y_\varepsilon}(x)| \leq 1$$

while, after extracting a subsequence if needed,

$$\mathbf{1}_{\varepsilon^{D-1}\tau_\varepsilon(x/\varepsilon^{D-1},v)>t} \rightharpoonup \Psi(t,v) \text{ in } L^\infty(\mathbf{R}_+ \times \mathbf{T}^D \times \mathbf{S}^{D-1}) \text{ weak-}*.$$

Therefore, if f is a weak-$*$ limit point of f_ε in $L^\infty(\mathbf{R}_+ \times \mathbf{T}^D \times \mathbf{S}^{D-1})$ as $\varepsilon \to 0$,

$$f(t,x,v) \geq f^{\mathrm{in}}(x-tv)\Psi(t,v) \text{ for a.e. } (t,x,v).$$

Step 3: Using the lower bound on the distribution of τ_r

Denoting by dv the uniform measure on $\mathbf{S}^{D-1}$,

$$\frac{1}{|\mathbf{S}^{D-1}|}\iint_{\mathbf{T}^D \times \mathbf{S}^{D-1}} f(t,x,v)^2\,dx\,dv$$

$$\geq \frac{1}{|\mathbf{S}^{D-1}|}\iint_{\mathbf{T}^D \times \mathbf{S}^{D-1}} f^{\mathrm{in}}(x-tv)^2\Psi(t,v)^2\,dx\,dv$$

$$= \int_{\mathbf{T}^D} f^{\mathrm{in}}(y)^2\,dy\,\frac{1}{|\mathbf{S}^{D-1}|}\int_{\mathbf{S}^{D-1}}\Psi(t,v)^2\,dv$$

$$\geq \|f^{\mathrm{in}}\|^2_{L^2(\mathbf{T}^D)}\left(\frac{1}{|\mathbf{S}^{D-1}|}\int_{\mathbf{S}^{D-1}}\Psi(t,v)\,dv\right)^2$$

$$= \|f^{\mathrm{in}}\|^2_{L^2(\mathbf{T}^D)}\Phi(t)^2.$$

By the Bourgain–Golse–Wennberg lower bound on the distribution Φ of free path lengths,

$$\|f(t,\cdot,\cdot)\|_{L^2(\mathbf{T}^D \times \mathbf{S}^{D-1})} \geq \frac{C_D}{t}\|f^{\mathrm{in}}\|_{L^2(\mathbf{T}^D)}, \quad t > 1.$$

On the other hand, by the spectral gap estimate, if f is a solution of the linear Boltzmann equation, one has

$$\|f(t,\cdot,\cdot)\|_{L^2(\mathbf{T}^D \times \mathbf{S}^{D-1})} \leq \int_{\mathbf{T}^D} f^{\mathrm{in}}(y)\,dy + Ce^{-\gamma t}\|f^{\mathrm{in}}\|_{L^2(\mathbf{T}^D)}$$

so that

$$\frac{C_D}{t} \le \frac{\|f^{\text{in}}\|_{L^1(\mathbf{T}^D)}}{\|f^{\text{in}}\|_{L^2(\mathbf{T}^D)}} + Ce^{-\gamma t}$$

for each $t > 1$.

Step 4: Choice of initial data

Pick ρ to be a bump function supported near $x = 0$ and such that

$$\int \rho(x)^2 \, dx = 1.$$

Take f^{in} to be $x \mapsto \lambda^{D/2}\rho(\lambda x)$ periodicized, so that

$$\int_{\mathbf{T}^D} f^{\text{in}}(x)^2 \, dx = 1, \quad \text{while} \quad \int_{\mathbf{T}^D} f^{\text{in}}(y) \, dy = \lambda^{-D/2} \int \rho(x) \, dx.$$

For such initial data, the inequality above becomes

$$\frac{C_D}{t} \le \lambda^{-D/2} \int \rho(x) \, dx + Ce^{-\gamma t}.$$

Conclude by choosing λ so that

$$\lambda^{-D/2} \int \rho(x) \, dx < \sup_{t>1} \left(\frac{C_D}{t} - Ce^{-\gamma t} \right) > 0. \qquad \square$$

Remarks.

1) The same result (with the same proof) holds for any smooth obstacle shape
   included in a shell

   $$\{x \in \mathbf{R}^D \mid C\varepsilon^D < \text{dist}(x, \varepsilon^{D-1}\mathbf{Z}^D) < C'\varepsilon^D\}.$$

2) The same result (with the same proof) holds if the specular reflection boundary condition is replaced by more general boundary conditions, such as ab-
   sorption (partial or complete) of the particles at the boundary of the obstacles, diffuse reflection, or any convex combination of specular and diffuse
   reflection —in the classical kinetic theory of gases, such boundary conditons
   are known as "accommodation boundary conditions".

3) But introducing even the smallest amount of stochasticity in any periodic
   configuration of obstacles can again lead to a Boltzmann–Grad limit that is
   described by the Lorentz kinetic model.

Example (Wennberg–Ricci [37]). In space dimension 2, take obstacles that are disks of radius r centered at the vertices of the lattice $r^{1/(2-\eta)}\mathbf{Z}^2$, assuming that $0 < \eta < 1$. In this case, Santaló's formula suggests that the free-path lengths scale like $r^{\eta/(2-\eta)} \to 0$.

Suppose the obstacles are removed independently with large probability — specifically, with probability $p = 1 - r^{\eta/(2-\eta)}$. In that case, the Lorentz kinetic equation governs the 1-particle density in the Boltzmann–Grad limit as $r \to 0^+$.

Having explained why neither the Lorentz kinetic equation nor any linear Boltzmann equation can govern the Boltzmann–Grad limit of the periodic Lorentz gas, in the remaining part of these notes we build the tools used in the description of that limit.

2.6 Coding particle trajectories with continued fractions

With the Bourgain–Golse–Wennberg lower bound for the distribution of free path lengths in the periodic Lorentz gas, we have seen that the 1-particle phase space density is bounded below by a quantity that is incompatible with the spectral gap of any linear Boltzmann equation —in particular with the Lorentz kinetic equation.

In order to further analyze the Boltzmann–Grad limit of the periodic Lorentz gas, we cannot content ourselves with even more refined estimates on the distribution of free path lengths, but we need a convenient way to encode particle trajectories.

More precisely, the two following problems must be answered somehow:

First problem: For a particle leaving the surface of an obstacle in a given direction, find the position of its next collision with an obstacle.

Second problem: Average —in some sense to be defined— in order to eliminate the direction dependence.

From now on, our discussion is limited to the case of spatial dimension $D = 2$, as we shall use continued fractions, a tool particularly well adapted to understanding the rational approximation of real numbers. Treating the case of a space dimension $D > 2$ along the same lines would require a better understanding of *simultaneous* rational approximation of $D - 1$ real numbers (by $D - 1$ rational numbers with the same denominator), a notoriously more difficult problem.

We first introduce some basic geometrical objects used in coding particle trajectories.

The first such object is the notion of *impact parameter*.

For a particle with velocity $v \in \mathbf{S}^1$ located at the position x on the surface of an obstacle (disk of radius r), we define its impact parameter $h_r(x, v)$ by the formula

$$h_r(x, v) = \sin(\widehat{n_x, v}).$$

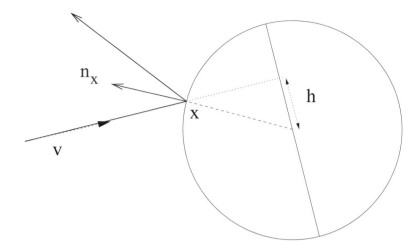

Figure 2.15: The impact parameter h corresponding with the collision point x at the surface of an obstacle, and a direction v

In other words, the absolute value of the impact parameter $h_r(x, v)$ is the distance of the center of the obstacle to the infinite line of direction v passing through x.

Obviously

$$h_r(x, \mathcal{R}[n_x]v) = h_r(x, v),$$

where we recall the notation $\mathcal{R}[n]v = v - 2(v \cdot n)n$.

The next important object in computing particle trajectories in the Lorentz gas is the *transfer map*.

For a particle leaving the surface of an obstacle in the direction v and with impact parameter h', define

$$T_r(h', v) = (s, h) \text{ with } \begin{cases} s = 2r \times \text{ distance to the next collision point,} \\ h = \text{ impact parameter at the next collision.} \end{cases}$$

Particle trajectories in the Lorentz gas are completely determined by the transfer map T_r and its iterates.

Therefore, a first step in finding the Boltzmann–Grad limit of the periodic, 2-dimensional Lorentz gas, is to compute the limit of T_r as $r \to 0^+$, in some sense that will be explained later.

At first sight, this seems to be a desperately hard problem to solve, as particle trajectories in the periodic Lorentz gas depend on their directions and the obstacle radius in the strongest possible way. Fortunately, there is an interesting property of rational approximation on the real line that greatly reduces the complexity of this problem.

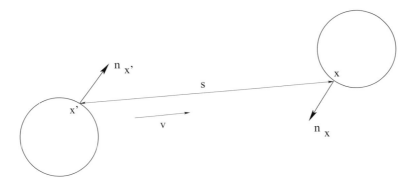

Figure 2.16· The transfer map

The 3-length theorem

Question (R. Thom, 1989). On a flat 2-torus with a disk removed, consider a linear flow with irrational slope. What is the longest orbit?

Theorem 2.6.1 (Blank–Krikorian [1]). *On a flat 2-torus with a segment removed, consider a linear flow with irrational slope $0 < \alpha < 1$. The orbits of this flow have at most three different lengths —exceptionally two, but generically three. Moreover, in the generic case where these orbits have exactly three different lengths, the length of the longest orbit is the sum of the two other lengths.*

These lengths are expressed in terms of the continued fraction expansion of the slope α.

Together with E. Caglioti in [9], we proposed the idea of using the Blank–Krikorian 3-length theorem to analyze particle paths in the 2-dimensional periodic Lorentz gas.

More precisely, orbits with the same lengths in the Blank–Krikorian theorem define a 3-term partition of the flat 2-torus into parallel strips, whose lengths and widths are computed exactly in terms of the continued fraction expansion of the slope (see Figure 2.17[1]).

The collision pattern for particles leaving the surface of one obstacle —and therefore the transfer map— can be explicitly determined in this way, for a.e. direction $v \in \mathbf{S}^1$.

In fact, there is a classical result known as the 3-length theorem, which is related to Blank–Krikorian's. Whereas the Blank–Krikorian theorem considers a linear flow with irrational slope on the flat 2-torus, the classical 3-length theorem is a statement about rotations of an irrational angle —i.e., about sections of the linear flow with irrational slope.

---

[1]Figures 2.16 and 2.17 are taken from a conference by E. Caglioti at the Centre International de Rencontres Mathématiques, Marseille, February 18–22, 2008.

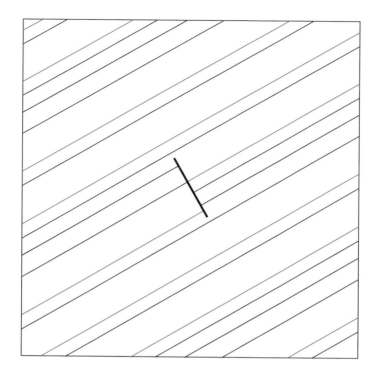

Figure 2.17: Three types of orbits: the blue orbit is the shortest, the red one is the longest, while the green one is of the intermediate length. The black segment removed is orthogonal to the direction of the trajectories.

Theorem 2.6.2 (3-length theorem). *Let $\alpha \in (0,1) \setminus \mathbf{Q}$ and $N \geq 1$. The sequence*

$$\{n\alpha \mid 0 \leq n \leq N\}$$

defines $N+1$ intervals on the circle of unit length $\simeq \mathbf{R}/\mathbf{Z}$. The lengths of these intervals take at most three different values.

This striking result was conjectured by H. Steinhaus, and proved in 1957 independently by P. Erdős, G. Hajos, J. Suranyi, N. Swieczkowski, P. Szüsz — reported in [42], and by Vera Sós [41].

As we shall see, the 3-length theorem (in either form) is the key to encoding particle paths in the 2-dimensional Lorentz gas. We shall need explicitly the formulas giving the lengths and widths of the three strips in the partition of the flat 2-torus defined by the Blank–Krikorian theorem. As this is based on the continued fraction expansion of the slope of the linear flow considered in the Blank–Krikorian theorem, we first recall some basic facts about continued fractions. An excellent reference for more information on this subject is [28].

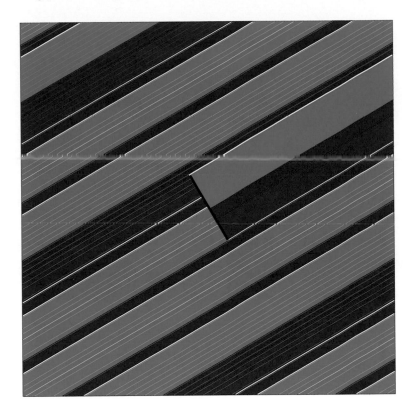

Figure 2.18: The 3-term partition. The shortest orbits are collected in the blue strip, the longest orbits in the red strip, while the orbits of intermediate length are collected in the green strip.

Continued fractions

Assume $0 < v_2 < v_1$ and set $\alpha = v_2/v_1$, and consider the continued fraction expansion of α:

$$\alpha = [0; a_0, a_1, a_2, \ldots] = \cfrac{1}{a_0 + \cfrac{1}{a_1 + \ldots}}.$$

Define the sequences of convergents $(p_n, q_n)_{n \geq 0}$, meaning that

$$\frac{p_{n+2}}{q_{n+2}} = [0; a_0, \ldots, a_n], \quad n \geq 2,$$

by the recursion formulas

$$p_{n+1} = a_n p_n + p_{n-1}, \qquad p_0 = 1, \ p_1 = 0,$$
$$q_{n+1} = a_n q_n + q_{n-1}, \qquad q_0 = 0, \ q_1 = 1.$$

Figure 2.19: Left: Hugo D. Steinhaus (1887–1972); right: Vera T. Sós

Finally, let d_n denote the sequence of errors

$$d_n = |q_n\alpha - p_n| = (-1)^{n-1}(q_n\alpha - p_n), \quad n \geq 0,$$

so that

$$d_{n+1} = -a_n d_n + d_{n-1}, \quad d_0 = 1, \; d_1 = \alpha.$$

The sequence d_n is decreasing and converges to 0, at least exponentially fast. (In fact, the irrational number for which the rational approximation by continued fractions is the slowest is the one for which the sequence of denominators q_n have the slowest growth, i.e., the golden mean

$$\theta = [0; 1, 1, \ldots] = \cfrac{1}{1 + \cfrac{1}{1 + \ldots}} = \frac{\sqrt{5} - 1}{2}.$$

The sequence of errors associated with θ satisfies $d_{n+1} = -d_n + d_{n-1}$ for each $n \geq 1$ with $d_0 = 1$ and $d_1 = \theta$, so that $d_n = \theta^n$ for each $n \geq 0$.)

By induction, one verifies that

$$q_n d_{n+1} + q_{n+1} d_n = 1, \quad n \geq 0.$$

Notation. we write $p_n(\alpha)$, $q_n(\alpha)$, $d_n(\alpha)$ to indicate the dependence of these quantities in α.

Collision patterns

The Blank–Krikorian 3-length theorem has the following consequence, of funda-mental importance in our analysis.

Any particle leaving the surface of one obstacle in some irrational direction v will next collide with one of *at most three* —exceptionally two— other obstacles.

Any such collision pattern involving the three obstacles seen by the depart-ing particle in the direction of its velocity is completely determined by exactly 4 parameters, computed in terms of the continued fraction expansion of v_2/v_1 —in the case where $0 < v_2 < v_1$, to which the general case can be reduced by obvious symmetry arguments.

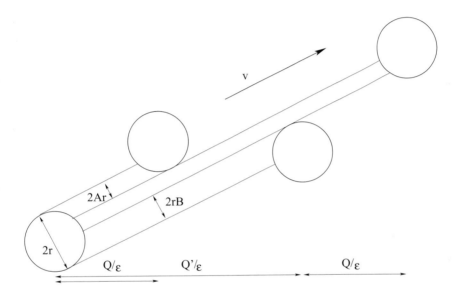

Figure 2.20: Collision pattern seen from the surface of one obstacle; $\varepsilon = 2r/v_1$

Assume therefore $0 < v_2 < v_1$ with $\alpha = v_2/v_1 \notin \mathbf{Q}$. Henceforth, we set $\varepsilon = 2r\sqrt{1+\alpha^2}$ and define

$$N(\alpha, \varepsilon) = \inf\{n \geq 0 \mid d_n(\alpha) \leq \varepsilon\},$$

$$k(\alpha, \varepsilon) = -\left[\frac{\varepsilon - d_{N(\alpha,\varepsilon)-1}(\alpha)}{d_{N(\alpha,\varepsilon)}(\alpha)}\right].$$

The parameters defining the collision pattern are A, B, Q —as they appear on the previous figure— together with an extra parameter $\Sigma \in \{\pm 1\}$. Here is how they are computed in terms of the continued fraction expansion of $\alpha = v_2/v_1$:

$$A(v,r) = 1 - \frac{d_{N(\alpha,\varepsilon)}(\alpha)}{\varepsilon},$$

$$B(v,r) = 1 - \frac{d_{N(\alpha,\varepsilon)-1}(\alpha)}{\varepsilon} + \frac{k(\alpha,\varepsilon)d_{N(\alpha,\varepsilon)}(\alpha)}{\varepsilon},$$

$$Q(v,r) = \varepsilon q_{N(\alpha,\varepsilon)}(\alpha),$$

$$\Sigma(v,r) = (-1)^{N(\alpha,\varepsilon)}.$$

The extra-parameter Σ in the list above has the following geometrical meaning. It determines the relative position of the closest and next to closest obstacles seen from the particle leaving the surface of the obstacle at the origin in the direction v.

The case represented on the figure where the closest obstacle is on top of the strip consisting of the longest particle path corresponds with $\Sigma = +1$; the case where that obstacle is at the bottom of this same strip corresponds with $\Sigma = -1$.

The figure above showing one example of collision pattern involves still another parameter, denoted by Q' on that figure.

This parameter Q' is not independent from A, B, Q, since one must have

$$AQ + BQ' + (1 - A - B)(Q + Q') = 1,$$

each term in this sum corresponding to the surface of one of the three strips in the 3-term partition of the 2-torus. (Remember that the length of the longest orbit in the Blank–Krikorian theorem is the sum of the two other lengths.) Therefore,

$$Q'(v,r) = \frac{1 - Q(v,r)(1 - B(v,r))}{1 - A(v,r)}.$$

Once the structure of collision patterns elucidated with the help of the Blank–Krikorian variant of the 3-length theorem, we return to our original problem, namely that of computing the transfer map.

In the next proposition, we shall see that the transfer map in a given, irrational direction $v \in \mathbf{S}^1$ can be expressed explicitly in terms of the parameters A, B, Q, Σ defining the collision pattern corresponding with this direction. Write

$$\mathbf{K} = (0,1)^3 \times \{\pm 1\}$$

and let $(A, B, Q, \Sigma) \in \mathbf{K}$ be the parameters defining the collision pattern seen by a particle leaving the surface of one obstacle in the direction v. Set

$$\mathbf{T}_{A,B,Q,\Sigma}(h') = \begin{cases} (Q, h' - 2\Sigma(1 - A)) & \text{if } 1 - 2A < \Sigma h' \leq 1, \\ (Q', h' + 2\Sigma(1 - B)) & \text{if } -1 \leq \Sigma h' < -1 + 2B, \\ (Q' + Q, h' + 2\Sigma(A - B)) & \text{if } -1 + 2B \leq \Sigma h' \leq 1 - 2A. \end{cases}$$

With this notation, the transfer map is essentially given by the explicit formula $\mathbf{T}_{A,B,Q,\Sigma}$, except for an error of the order $O(r^2)$ on the free-path length from obstacle to obstacle.

Proposition 2.6.3 (Caglioti–Golse [10, 11]). *One has*

$$T_r(h', v) = \mathbf{T}_{(A,B,Q,\Sigma)(v,r)}(h') + (O(r^2), 0)$$

in the limit as $r \to 0^+$.

In fact, the proof of this proposition can be read on the figure above that represents a generic collision pattern. The first component in the explicit formula

$$\mathbf{T}_{(A,B,Q,\Sigma)(v,r)}(h')$$

represents exactly $2r$ times the distance between the vertical segments that are the projections of the diameters of the 4 obstacles on the vertical ordinate axis. Obviously, the free-path length from obstacle to obstacle is the distance between the corresponding vertical segments, minus a quantity of the order $O(r)$ that is the distance from the surface of the obstacle to the corresponding vertical segment.

On the other hand, the second component in the same explicit formula is exact, as it relates impact parameters, which are precisely the intersections of the infinite line that contains the particle path with the vertical segments corresponding with the two obstacles joined by this particle path.

If we summarize what we have done so far, we see that we have solved our first problem stated at the beginning of the present section, namely that of finding a convenient way of coding the billiard flow in the periodic case and for space dimension 2, for a.e. given direction v.

2.7 An ergodic theorem for collision patterns

It remains to solve the second problem, namely, to find a convenient way of averaging the computation above so as to get rid of the dependence on the direction v.

Before going further in this direction, we need to recall some known facts about the ergodic theory of continued fractions.

The Gauss map

Consider the Gauss map, which is defined on all irrational numbers in $(0, 1)$ as follows:

$$T : (0, 1) \setminus \mathbf{Q} \ni x \longmapsto Tx = \tfrac{1}{x} - \left[\tfrac{1}{x}\right] \in (0, 1) \setminus \mathbf{Q}.$$

This Gauss map has the following invariant probability measure —found by Gauss himself:

$$dg(x) = \frac{1}{\ln 2} \frac{dx}{1 + x}.$$

Moreover, the Gauss map T is ergodic for the invariant measure $dg(x)$. By Birkhoff's theorem, for each $f \in L^1(0,1;dg)$,

$$\frac{1}{N} \sum_{k=0}^{N-1} f(T^k x) \longrightarrow \int_0^1 f(z)\, dg(z) \text{ a.e. in } x \in (0,1)$$

as $N \to +\infty$.

How the Gauss map is related to continued fractions is explained as follows: for

$$\alpha = [0; a_0, a_1, a_2, \ldots] = \cfrac{1}{a_0 + \cfrac{1}{a_1 + \ldots}} \in (0,1) \setminus \mathbf{Q},$$

the terms $a_k(\alpha)$ of the continued fraction expansion of α can be computed from the iterates of the Gauss map acting on α. Specifically,

$$a_k(\alpha) = \left[\frac{1}{T^k \alpha}\right], \quad k \geq 0.$$

As a consequence, the Gauss map corresponds with the shift to the left on infinite sequences of positive integers arising in the continued fraction expansion of irrationals in $(0,1)$. In other words,

$$T[0; a_0, a_1, a_2, \ldots] = [0; a_1, a_2, a_3 \ldots],$$

equivalently recast as

$$a_n(T\alpha) = a_{n+1}(\alpha), \quad n \geq 0.$$

Thus, the terms $a_k(\alpha)$ of the continued fraction expansion of any $\alpha \in (0,1)\setminus\mathbf{Q}$ are easily expressed in terms of the sequence of iterates $(T^k\alpha)_{k\geq 0}$ of the Gauss map acting on α. The error $d_n(\alpha)$ is also expressed in terms of that same sequence $(T^k\alpha)_{k\geq 0}$, by equally simple formulas.

Starting from the induction relation on the error terms

$$d_{n+1}(\alpha) = -a_n(\alpha)d_n(\alpha) + d_{n-1}(\alpha), \quad d_0(\alpha) = 1, \ d_1(\alpha) = \alpha,$$

and the explicit formula relating $a_n(T\alpha)$ to $a_n(\alpha)$, we see that

$$\alpha d_n(T\alpha) = d_{n+1}(\alpha), \quad n \geq 0.$$

This entails the formula

$$d_n(\alpha) = \prod_{k=0}^{n-1} T^k \alpha, \quad n \geq 0.$$

Observe that, for each $\theta \in [0,1] \setminus \mathbf{Q}$, one has

$$\theta \cdot T\theta < \tfrac{1}{2},$$

so that

$$d_n(\alpha) \leq 2^{-[n/2]}, \quad n \geq 0,$$

which establishes the exponential decay mentioned above. (As a matter of fact, exponential convergence is the slowest possible for the continued fraction algorithm, as it corresponds with the rational approximation of algebraic numbers of degree 2, which are the hardest to approximate by rational numbers.)

Unfortunately, the dependence of $q_n(\alpha)$ in α is more complicated. Yet one can find a way around this, with the following observation. Starting from the relation

$$q_{n+1}(\alpha)d_n(\alpha) + q_n(\alpha)d_{n+1}(\alpha) = 1,$$

we see that

$$q_n(\alpha)d_{n-1}(\alpha) = \sum_{j=1}^{n}(-1)^{n-j}\frac{d_n(\alpha)d_{n-1}(\alpha)}{d_j(\alpha)d_{j-1}(\alpha)}$$

$$= \sum_{j=1}^{n}(-1)^{n-j}\prod_{k=j}^{n-1}T^{k-1}\alpha T^k\alpha.$$

Using once more the inequality $\theta \cdot T\theta < \frac{1}{2}$ for $\theta \in [0,1] \setminus \mathbf{Q}$, one can truncate the summation above at the cost of some exponentially small error term. Specifically, one finds that

$$\left| q_n(\alpha)d_{n-1}(\alpha) - \sum_{j=n-l}^{n}(-1)^{n-j}\frac{d_n(\alpha)d_{n-1}(\alpha)}{d_j(\alpha)d_{j-1}(\alpha)} \right|$$

$$= \left| q_n(\alpha)d_{n-1}(\alpha) - \sum_{j=n-l}^{n}(-1)^{n-j}\prod_{k=j}^{n-1}T^{k-1}\alpha T^k\alpha \right| \leq 2^{-l}.$$

More information on the ergodic theory of continued fractions can be found in the classical monograph [28] on continued fractions, and in Sinai's book on ergodic theory [40].

An ergodic theorem

We have seen in the previous section that the transfer map satisfies

$$T_r(h',v) = \mathbf{T}_{(A,B,Q,\Sigma)(v,r)}(h') + (O(r^2),0) \text{ as } r \to 0^+$$

for each $v \in \mathbf{S}^1$ such that $v_2/v_1 \in (0,1) \setminus \mathbf{Q}$.

Obviously, the parameters (A,B,Q,Σ) are extremely sensitive to variations in v and r as $r \to 0^+$, so that even the explicit formula for $T_{A,B,Q,\Sigma}$ is not too useful in itself.

Each time one must handle a strongly oscillating quantity such as the free path length $\tau_r(x,v)$ or the transfer map $T_r(h',v)$, it is usually a good idea to consider the distribution of that quantity under some natural probability measure rather than the quantity itself. Following this principle, we are led to consider the family of probability measures in $(s, h) \in \mathbf{R}_+ \times [-1, 1]$,

$$\delta((s, h) - T_r(h', v)),$$

or equivalently

$$\delta((s, h) - T_{(A,B,Q,\Sigma)(v,r)}(h')).$$

A first obvious idea would be to average out the dependence in v of this family of measures: as we shall see later, this is not an easy task.

A somewhat less obvious idea is to average over obstacle radius. Perhaps surprisingly, this is easier than averaging over the direction v.

That averaging over obstacle radius is a natural operation in this context can be explained by the following observation. We recall that the sequence of errors $d_n(\alpha)$ in the continued fraction expansion of an irrational $\alpha \in (0, 1)$ satisfies

$$\alpha d_n(T\alpha) = d_{n+1}(\alpha), \quad n \geq 0,$$

so that

$$N(\alpha, \varepsilon) = \inf\{n \geq 1 \mid d_n(\alpha) \leq \varepsilon\}$$

is transformed by the Gauss map as follows:

$$N(a, \varepsilon) = N(T\alpha, \varepsilon/\alpha) + 1.$$

In other words, the transfer map for the 2-dimensional periodic Lorentz gas in the billiard table Z_r (meaning with circular obstacles of radius r centered at the vertices of the lattice $\mathbf{Z}^2$) in the direction v corresponding with the slope α is essentially the same as for the billiard table $Z_{r/\alpha}$ but in the direction corresponding with the slope $T\alpha$. Since the problem is invariant under the transformation

$$\alpha \mapsto T\alpha, \qquad r \mapsto r/\alpha,$$

this suggests the idea of averaging with respect to the scale invariant measure in the variable r, i.e., dr/r on $\mathbf{R}_+^*$.

The key result in this direction is the following ergodic lemma for functions that depend on *finitely many* d_ns.

Lemma 2.7.1 (Caglioti–Golse [9, 22, 11]). *For $\alpha \in (0, 1) \setminus \mathbf{Q}$, set*

$$N(\alpha, \varepsilon) = \inf\{n \geq 0 \mid d_n(\alpha) \leq \varepsilon\}.$$

For each $m \geq 0$ and each $f \in C(\mathbf{R}_+^{m+1})$, one has

$$\frac{1}{|\ln \eta|} \int_\eta^{1/4} f\left(\frac{d_{N(\alpha,\varepsilon)}(\alpha)}{\varepsilon}, \dots, \frac{d_{N(\alpha,\varepsilon)-m}(\alpha)}{\varepsilon} \right) \frac{d\varepsilon}{\varepsilon} \longrightarrow L_m(f)$$

a.e. in $\alpha \in (0, 1)$ as $\eta \to 0^+$, where the limit $L_m(f)$ is independent of α.

With this lemma, we can average over obstacle radius any function that depends on collision patterns, i.e., any function of the parameters A, B, Q, Σ.

Proposition 2.7.2 (Caglioti–Golse [11]). *Let* $\mathbf{K} = [0, 1]^3 \times \{\pm 1\}$. *For each* $F \in C(\mathbf{K})$, *there exists* $\mathcal{L}(F) \in \mathbf{R}$ *independent of* v *such that*

$$\frac{1}{\ln(1/\eta)} \int_\eta^{1/2} F(A(v, r), B(v, r), Q(v, r), \Sigma(v, r)) \frac{dr}{r} \longrightarrow \mathcal{L}(F)$$

for a.e. $v \in \mathbf{S}^1$ *such that* $0 < v_2 < v_1$ *in the limit as* $\eta \to 0^+$.

Sketch of the proof. First eliminate the Σ dependence by decomposing

$$F(A, B, Q, \Sigma) = F_+(A, B, Q) + \Sigma F_-(A, B, Q).$$

Hence it suffices to consider the case where $F \equiv F(A, B, Q)$.

Setting $\alpha = v_2/v_1$ and $\varepsilon = 2r/v_1$, we recall that

$$A(v, r) \text{ is a function of } \frac{d_{N(\alpha,\varepsilon)}(\alpha)}{\varepsilon},$$

$$B(v, r) \text{ is a function of } \frac{d_{N(\alpha,\varepsilon)}(\alpha)}{\varepsilon} \text{ and } \frac{d_{N(\alpha,\varepsilon)-1}(\alpha)}{\varepsilon}.$$

As for the dependence of F on Q, proceed as follows: in $F(A, B, Q)$, replace $Q(v, r)$ with

$$\frac{\varepsilon}{d_{N(\alpha,\varepsilon)-1}} \sum_{j=N(\alpha,\varepsilon)-m}^{N(\alpha,\varepsilon)} (-1)^{N(\alpha,\varepsilon)-j} \frac{d_{N(\alpha,\varepsilon)}(\alpha) d_{N(\alpha,\varepsilon)-1}(\alpha)}{d_j(\alpha) d_{j-1}(\alpha)},$$

at the expense of an error term of the order

$$O(\text{modulus of continuity of } F(2^{-m})) \to 0 \text{ as } m \to \infty,$$

uniformly as $\varepsilon \to 0^+$.

This substitution leads to an integrand of the form

$$f\left(\frac{d_{N(\alpha,\varepsilon)}(\alpha)}{\varepsilon}, \dots, \frac{d_{N(\alpha,\varepsilon)-m-1}(\alpha)}{\varepsilon}\right)$$

to which we apply the ergodic lemma above: its Cesàro mean converges, in the small radius limit, to some limit $\mathcal{L}_m(F)$ independent of α.

By uniform continuity of F, one finds that

$$|\mathcal{L}_m(F) - \mathcal{L}_{m'}(F)| = O(\text{modulus of continuity of } F(2^{-m \wedge m'}))$$

(with the notation $m \wedge m' = \min(m, m')$), so that $\mathcal{L}_m(F)$ is a Cauchy sequence as $m \to \infty$. Hence

$$\mathcal{L}_m(F) \to \mathcal{L}(F) \text{ as } m \to \infty$$

and with the error estimate above for the integrand, one finds that

$$\frac{1}{\ln(1/\eta)} \int_\eta^{1/2} F(A(v,r), B(v,r), Q(v,r), \Sigma(v,r)) \frac{dr}{r} \longrightarrow \mathcal{L}(F)$$

as $\eta \to 0^+$.   □

With the ergodic theorem above, and the explicit approximation of the transfer map expressed in terms of the parameters (A, B, Q, Σ) that determine collision patterns in any given direction v, we easily arrive at the following notion of a "probability of transition" for a particle leaving the surface of an obstacle with an impact parameter h' to hit the next obstacle on its trajectory at time s/r with an impact parameter h.

Theorem 2.7.3 (Caglioti–Golse [10, 11]). *For each $h' \in [-1, 1]$, there exists a probability density $P(s, h|h')$ on $\mathbf{R}_+ \times [-1, 1]$ such that, for each f belonging to $C(\mathbf{R}_+ \times [-1, 1])$,*

$$\frac{1}{|\ln\eta|} \int_\eta^{1/4} f(T_r(h', v)) \frac{dr}{r} \longrightarrow \int_0^\infty \int_{-1}^1 f(s, h) P(s, h|h') \, ds \, dh$$

a.e. in $v \in \mathbf{S}^1$ as $\eta \to 0^+$.

In other words, the transfer map converges in distribution and in the sense of Cesàro, in the small radius limit, to a transition probability $P(s, h|h')$ that is independent of v.

We are therefore left with the following problems:

a) to compute the transition probability $P(s, h|h')$ explicitly and discuss its properties, and

b) to explain the role of this transition probability in the Boltzmann–Grad limit of the periodic Lorentz gas dynamics.

2.8 Explicit computation of the transition probability $P(s, h|h')$

Most unfortunately, our argument leading to the existence of the limit $\mathcal{L}(F)$, the core result of the previous section, cannot be used for computing explicitly the value $\mathcal{L}(F)$. Indeed, the convergence proof is based on the ergodic lemma in the last section, coupled to a sequence of approximations of the parameter Q in collision patterns that involve only finitely many error terms $d_n(\alpha)$ in the continued fraction expansion of α. The existence of the limit is obtained through Cauchy's criterion, precisely because of the difficulty in finding an explicit expression for the limit.

Nevertheless, we have arrived at the following expression for the transition probability $P(s, h|h')$:

Theorem 2.8.1 (Caglioti–Golse [10, 11]). *The transition density* $P(s, h|h')$ *is expressed in terms of* $a = \frac{1}{2}|h - h'|$ *and* $b = \frac{1}{2}|h + h'|$ *by the explicit formula*

$$P(s, h|h') = \frac{3}{\pi^2 sa}\Big[\big((s - \tfrac{1}{2}sa) \wedge (1 + \tfrac{1}{2}sa) - 1 \vee (\tfrac{1}{2}s + \tfrac{1}{2}sb)\big)_+$$
$$+ \big((s - \tfrac{1}{2}sa) \wedge 1 - (\tfrac{1}{2}s + \tfrac{1}{2}sb) \vee \big(1 - \tfrac{1}{2}sa\big)\big)_+$$
$$+ sa \wedge |1 - s|\,\mathbf{1}_{s<1} + (sa - |1 - s|)_+\Big],$$

with the notations $x \wedge y = \min(x, y)$ *and* $x \vee y = \max(x, y)$.
    Moreover, the function

$$(s, h, h') \mapsto (1 + s)P(s, h|h') \text{ belongs to } L^2(\mathbf{R}_+ \times [-1, 1]^2).$$

    In fact, the key result in the proof of this theorem is the asymptotic distribution of 3-obstacle collision patterns —i.e., the computation of the limit $\mathcal{L}(f)$, whose existence has been proved in the last section's proposition.

Theorem 2.8.2 (Caglioti–Golse [11]). *Define* $\mathbf{K} = [0, 1]^3 \times \{\pm 1\}$. *Then, for each* $F \in C(\mathbf{K})$,

$$\frac{1}{|\ln \eta|} \int_\eta^{1/4} F((A, B, Q, \Sigma)(v, r))\frac{dr}{r} \longrightarrow \mathcal{L}(F)$$
$$= \int_\mathbf{K} F(A, B, Q, \Sigma)\, dm(A, B, Q, \Sigma) \text{ a.e. in } v \in \mathbf{S}^1$$

as $\eta \to 0^+$, *where*

$$dm(A, B, Q, \Sigma) = dm_0(A, B, Q) \otimes \tfrac{1}{2}(\delta_{\Sigma=1} + \delta_{\Sigma=-1}),$$
$$dm_0(A, B, Q) = \frac{12}{\pi^2}\,\mathbf{1}_{0<A<1}\,\mathbf{1}_{0<B<1-A}\,\mathbf{1}_{0<Q<\frac{1}{2-A-B}}\,\frac{dA\,dB\,dQ}{1 - A}.$$

    Before giving an idea of the proof of the theorem above on the distribution of 3-obstacle collision patterns, it is perhaps worthwhile explaining why the measure m above is somehow natural in the present context.

    To begin with, the constraints $0 < A < 1$ and $0 < B < 1 - A$ have an obvious geometric meaning (see Figure 2.20 on collision patterns.) More precisely, the widths of the three strips in the 3-term partition of the 2-torus minus the slit constructed in the penultimate section (as a consequence of the Blank–Krikorian 3-length theorem) add up to 1. Since A is the width of the strip consisting of the shortest orbits in the Blank–Krikorian theorem, and B that of the strip consisting of the next to shortest orbits, one has

$$0 < A + B \leq 1$$

with equality only in the exceptional case where the orbits have at most two different lengths, which occurs for a set of measure 0 in v or r. Therefore, one has

$$0 < B(v, r) < 1 - A(v, r), \quad \text{for a.e. } r \in (0, \tfrac{1}{2}).$$

Likewise, the total area of the 2-torus is the sum of the areas of the strips consisting of all orbits with the three possible lengths:

$$1 = QA + Q'B + (Q + Q')(1 - A - B) = Q(1 - B) + Q'(1 - A)$$
$$\geq Q(2 - A - B)$$

as $Q' \geq Q$ (see again the figure above on collision patterns).

Therefore, the volume element

$$\frac{dA\,dB\,dQ}{1 - A}$$

in the expression of dm_0 implies that the parameters A, $\frac{B}{1-A}$ —or equivalently B mod $(1 - A)$— and Q are uniformly distributed in the largest subdomain of $[0, 1]^3$ that is compatible with the geometric constraints.

The first theorem is a consequence of the second: indeed, $P(s, h|h')\,ds\,dh$ is the image measure of $dm(A, B, Q, \Sigma)$ under the map

$$\mathbf{K} \ni (A, B, Q, \Sigma) \longmapsto T_{(A,B,Q,\Sigma)}(h', v).$$

That $(1 + s)P(s, h|h')$ is square integrable is proved by inspection —by using the explicit formula for $P(s, h|h')$.

Therefore, it remains to prove the second theorem.

We are first going to show that the family of averages over velocities satisfy

$$\int_{\substack{|v|=1 \\ 0 < v_2 < v_1}} F(A(v, r), B(v, r), Q(v, r), \Sigma(v, r))\,dv$$

$$\rightarrow \frac{\pi}{8} \int_{\mathbf{K}} F(A, B, Q, \Sigma)\,dm(A, B, Q, \Sigma)$$

as $r \to 0^+$ for each $F \in C_b(\mathbf{K})$.

On the other hand, because of the proposition in the previous section,

$$\frac{1}{\ln(1/\eta)} \int_\eta^{1/2} F(A(v, r), B(v, r), Q(v, r), \Sigma(v, r))\frac{dr}{r} \longrightarrow \mathcal{L}(F)$$

for a.e. $v \in \mathbf{S}^1$ such that $0 < v_2 < v_1$ in the limit as $\eta \to 0^+$.

Since we know that the limit $\mathcal{L}(F)$ is independent of v, comparing the two convergence statements above shows that

$$\mathcal{L}(F) = \int_{\mathbf{K}} F(A, B, Q, \Sigma)\,dm(A, B, Q, \Sigma).$$

Therefore, we are left with the task of computing

$$\lim_{r \to 0^+} \int_{\substack{|v|=1 \\ 0 < v_2 < v_1}} F(A(v, r), B(v, r), Q(v, r), \Sigma(v, r))\,dv.$$

The method for computing this type of expression is based on

a) Farey fractions (sometimes called "slow continued fractions"), and

b) estimates for Kloosterman's sums, due to Boca–Zaharescu [3].

To begin with, we need to recall a few basic facts about Farey fractions.

Farey fractions

Put a filtration on the set of rationals in $[0, 1]$ as follows:

$$\mathcal{F}_{\mathcal{Q}} = \left\{ \tfrac{p}{q} \mid 0 \le p \le q \le \mathcal{Q}, \ \gcd(p, q) = 1 \right\},$$

indexed in increasing order.

$$0 = \frac{0}{1} < \gamma_1 < \cdots < \gamma_j = \frac{p_j}{q_j} < \cdots < \gamma_{\varphi(\mathcal{Q})} = \frac{1}{1} = 1,$$

where φ denotes Euler's totient function:

$$\phi(n) = n \prod_{\substack{p \text{ prime} \\ p \mid n}} \left(1 - \frac{1}{p} \right).$$

An important operation in the construction of Farey fractions is the notion of "mediant" of two fractions. Given two rationals

$$\gamma = \frac{p}{q} \quad \text{and} \quad \hat{\gamma} = \frac{\hat{p}}{\hat{q}}$$

with $0 \le p \le q$, $0 \le \hat{p} \le \hat{q}$, and $\gcd(p, q) = \gcd(\hat{p}, \hat{q}) = 1$, their *mediant* is defined as

$$\text{mediant} = \gamma \oplus \hat{\gamma} = \frac{p + \hat{p}}{q + \hat{q}} \in (\gamma, \hat{\gamma}).$$

Hence, if $\gamma = \frac{p}{q} < \hat{\gamma} = \frac{\hat{p}}{\hat{q}}$ are adjacent in $\mathcal{F}_{\mathcal{Q}}$, then

$$\hat{a}q - a\hat{q} = 1 \quad \text{and} \quad q + \hat{q} > \mathcal{Q}.$$

Conversely, $q, \hat{q}$ are denominators of adjacent fractions in $\mathcal{F}_{\mathcal{Q}}$ if and only if

$$0 \le q, \hat{q} \le \mathcal{Q}, \quad q + \hat{q} > \mathcal{Q}, \quad \gcd(q, q') = 1.$$

Given $\alpha \in (0, 1) \setminus \mathbf{Q}$ and $\mathcal{Q} \ge 1$, there exists a unique pair of adjacent Farey fractions in $\mathcal{F}_{\mathcal{Q}}$, henceforth denoted $\gamma(\alpha, \mathcal{Q})$ and $\hat{\gamma}(\alpha, \mathcal{Q})$, such that

$$\gamma(\alpha, \mathcal{Q}) = \frac{p(\alpha, \mathcal{Q})}{q(\alpha, \mathcal{Q})} < \alpha < \hat{\gamma}(\alpha, \mathcal{Q}) = \frac{\hat{p}(\alpha, \mathcal{Q})}{\hat{q}(\alpha, \mathcal{Q})}.$$

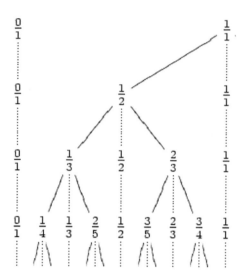

Figure 2.21: The Stern–Brocot tree. Each fraction γ on the n-th line is the mediant of the two fractions closest to γ on the $(n-1)$-st line. The first line consists of 0 and 1 written as $0 = \frac{0}{1}$ and $1 = \frac{1}{1}$. Each rational in $[0,1]$ is obtained in this way.

At this point, we recall the relation between Farey and continued fractions. Pick $0 < \varepsilon < 1$; we recall that, for each $\alpha \in (0,1) \setminus \mathbf{Q}$,

$$N(\alpha, \varepsilon) = \min\{n \in \mathbf{N} \mid d_n(\alpha) \leq \varepsilon\}, \quad d_n(\alpha) = \mathrm{dist}(q_n(\alpha)\alpha, \mathbf{Z}).$$

Set $\mathcal{Q} = [1/\varepsilon]$, and let

$$\gamma(\alpha, \mathcal{Q}) = \frac{p(\alpha, \mathcal{Q}))}{q(\alpha, \mathcal{Q})} < \hat{\gamma}(\alpha, \mathcal{Q}) = \frac{\hat{p}(\alpha, \mathcal{Q}))}{\hat{q}(\alpha, \mathcal{Q})}$$

with $\gcd(p(\alpha, \mathcal{Q}), q(\alpha, \mathcal{Q})) = \gcd(\hat{p}(\alpha, \mathcal{Q}), \hat{q}(\alpha, \mathcal{Q})) = 1$ be the two adjacent Farey fractions in $\mathcal{F}_{\mathcal{Q}}$ surrounding α. Then:

a) one of the integers $q(\alpha, \mathcal{Q})$ and $\hat{q}(\alpha, \mathcal{Q})$ is the denominator of the $N(\alpha, \varepsilon)$-th convergent in the continued fraction expansion of α, i.e., $q_{N(\alpha, \varepsilon)}(\alpha)$, and

b) the other is of the form

$$m q_{N(\alpha, \varepsilon)} + q_{N(\alpha, \varepsilon)-1}, \quad \text{with } 0 \leq m \leq a_{N(\alpha, \varepsilon)}(\alpha),$$

where we recall that

$$\alpha = [0; a_1, a_2, \ldots] = \cfrac{1}{a_0 + \cfrac{1}{a_1 + \ldots}}.$$

Setting $\alpha = v_2/v_1$ and $\varepsilon = 2r/v_1$, we recall that, by definition

$$Q(v, r) = \varepsilon q_{N(\alpha, \varepsilon)}(\alpha) \in \{\varepsilon q(\alpha, \mathcal{Q}), \varepsilon \hat{q}(\alpha, \mathcal{Q})\} \text{ with } \mathcal{Q} = [1/\varepsilon],$$

and we further define

$$D(v, r) = d_{N(\alpha, \varepsilon)}/\varepsilon = \text{dist}(\tfrac{1}{\varepsilon}Q(v, r)\alpha, \mathbf{Z})/\varepsilon,$$

and

$$\tilde{Q}(v, r) = \begin{cases} \varepsilon \hat{q}(\alpha, \mathcal{Q}) \text{ if } q_{N(\alpha, L)}(\alpha) = q(\alpha, \mathcal{Q}), \\ \varepsilon q(\alpha, \mathcal{Q}) \text{ if } q_{N(\alpha, \varepsilon)}(\alpha) = \hat{q}(\alpha, \mathcal{Q}). \end{cases}$$

Now, we recall that $A(v, r) = 1 - D(v, r)$; moreover, we see that

$$B(v, r) = 1 - \frac{d_{N(\alpha, \varepsilon)-1}(\alpha)}{\varepsilon} - \left[\frac{1 - d_{N(\alpha, \varepsilon)-1}(\alpha)/\varepsilon}{D(v, r)}\right] D(v, r)$$

$$= 1 - d_{N(\alpha, \varepsilon)-1}(\alpha)/\varepsilon \bmod D(v, r)$$

$$= 1 - \text{dist}(\tfrac{1}{\varepsilon}\tilde{Q}(v, r)\alpha, \mathbf{Z})/\varepsilon \bmod D(v, r).$$

To summarize, we have

$$F(A(v, r), B(v, r), Q(v, r)) = G(Q(v, r), \tilde{Q}(v, r), D(v, r))$$

and we are left with the task of computing

$$\lim_{r \to 0^+} \int_{\mathbf{S}^1_+} G(Q(v, r), \tilde{Q}(v, r), D(v, r))\, dv$$

where $\mathbf{S}^1_+$ is the first octant in the unit circle. The other octants in the unit circle give the same contribution by obvious symmetry arguments.

More specifically:

Lemma 2.8.3. *Let* $\alpha \in (0, 1) \setminus \mathbf{Q}$, *and let* $\frac{p}{q} < \alpha < \frac{\hat{p}}{\hat{q}}$ *be the two adjacent Farey fractions in* $\mathcal{F}_\mathcal{Q}$ *surrounding* α, *with* $\mathcal{Q} = [1/\varepsilon]$. *Then:*

a) *If* $\frac{p}{q} < \alpha \le \frac{\hat{p}-\varepsilon}{\hat{q}}$, *then*

$$Q(v, r) = \varepsilon q, \quad \tilde{Q}(v, r) = \varepsilon \hat{q}, \quad D(v, r) = \tfrac{1}{\varepsilon}(\alpha q - p).$$

b) *If* $\frac{p+\varepsilon}{q} < \alpha < \frac{\hat{p}}{\hat{q}}$, *then*

$$Q(v, r) = \varepsilon \hat{q}, \quad \tilde{Q}(v, r) = \varepsilon q, \quad D(v, r) = \tfrac{1}{\varepsilon}(\hat{p} - \alpha \hat{q}).$$

c) *If* $\frac{p+\varepsilon}{q} < \alpha \le \frac{\hat{p}-\varepsilon}{\hat{q}}$, *then*

$$Q(v, r) = \varepsilon(q \wedge \hat{q}), \quad \tilde{Q}(v, r) = \varepsilon(q \vee \hat{q}), \quad D(v, r) = \text{dist}(\tfrac{1}{\varepsilon}Q(v, r)\alpha, \mathbf{Z}).$$

Therefore, assuming for simplicity that

$$G(x, y, z) = g(x, y) H'(z) \text{ and } \varepsilon = 1/\mathcal{Q},$$

one has

$$\int_{\mathbf{S}^1_+} G(Q(v, r), \hat{Q}(v, r), D(v, r)) \, dv$$

$$= \sum_{\substack{0 < q, \hat{q} \le \mathcal{Q} < q + \hat{q} \\ \gcd(q, \hat{q}) = 1}} \int_{p/q}^{(\hat{p} - \varepsilon)/\hat{q}} g\left(\frac{q}{\mathcal{Q}}, \frac{\hat{q}}{\mathcal{Q}}\right) H'(\mathcal{Q}(q\alpha - p)) \, d\alpha$$

$$+ \text{ three other similar terms}$$

$$= \sum_{\substack{0 < q, \hat{q} \le \mathcal{Q} < q + \hat{q} \\ \gcd(q, \hat{q}) = 1}} g\left(\frac{q}{\mathcal{Q}}, \frac{\hat{q}}{\mathcal{Q}}\right) \frac{1}{q\mathcal{Q}} \left(H\left(\frac{1 - q/\mathcal{Q}}{\hat{q}/\mathcal{Q}}\right) - H(0) \right)$$

$$+ \text{ three other similar terms.}$$

Thus, everything reduces to computing

$$\lim_{\mathcal{Q} \to +\infty} \frac{1}{\mathcal{Q}^2} \sum_{\substack{0 < q, \hat{q} \le \mathcal{Q} < q + \hat{q} \\ \gcd(q, \hat{q}) = 1}} \psi\left(\frac{q}{\mathcal{Q}}, \frac{\hat{q}}{\mathcal{Q}}\right).$$

We conclude with the following

Lemma 2.8.4 (Boca–Zaharescu [3]). *For $\psi \in C_c(\mathbf{R}^2)$, one has*

$$\frac{1}{\mathcal{Q}^2} \sum_{\substack{0 < q, \hat{q} \le \mathcal{Q} < q + \hat{q} \\ \gcd(q, \hat{q}) = 1}} \psi\left(\frac{q}{\mathcal{Q}}, \frac{\hat{q}}{\mathcal{Q}}\right) \to \frac{6}{\pi^2} \iint_{0 < x, y < 1 < x + y} \psi(x, y) \, dx \, dy$$

in the limit as $\mathcal{Q} \to \infty$.

    This is precisely the path followed by F. Boca and A. Zaharescu to compute the limiting distribution of free path lengths in [3] (see Theorem 2.4.6); as explained above, their analysis can be greatly generalized in order to compute the transition probability that is the limit of the transfer map as the obstacle radius $r \to 0^+$.

2.9   A kinetic theory in extended phase-space for the Boltzmann–Grad limit of the periodic Lorentz gas

We are now ready to propose an equation for the Boltzmann–Grad limit of the periodic Lorentz gas in space dimension 2. For each $r \in (0, \frac{1}{2})$, denote the billiard map by

$$\mathcal{B}_r : \Gamma_r^+ \ni (x, v) \longmapsto \mathcal{B}_r(x, v) = (x + \tau_r(x, v)v, \mathcal{R}[x + \tau_r(x, v)v]v) \in \Gamma_r^+.$$

For $(x_0, v_0) \in \Gamma_r^+$, set

$$(x_n, v_n) = \mathcal{B}_r^n(x_0, v_0)$$

and define

$$b_r^n(x, v) = (A, B, Q, \Sigma)(v_n, r), \quad n \in \mathbf{N}^*.$$

Henceforth, for each $n \geq 1$, we write

$$\mathcal{K}_n = \mathbf{R}^2 \times \mathbf{S}^1 \times \mathbf{R}_+ \times [-1, 1] \times \mathbf{K}^n.$$

We make the following asymptotic independence hypothesis: there exists a probability measure Π on $\mathbf{R}_+ \times [-1, 1]$ such that, for each $n \geq 1$ and each Ψ in $C(\mathcal{K}_n)$ with compact support,

$$
(H) \quad
\begin{aligned}
&\lim_{r \to 0^+} \int_{Z_r \times \mathbf{S}^1} \Psi\left(x, v, r\tau_r\left(\tfrac{x}{r}, v\right), h_r\left(\tfrac{x_1}{r}, v_1\right), b_r^1, \dots, b_r^n\right) dx\, dv \\
&= \int_{Q_n} \Psi(x, v, \tau, h, \beta_1, \dots, \beta_n)\, dx\, dv\, d\Pi(\tau, h)\, dm(\beta_1) \dots dm(\beta_n),
\end{aligned}
$$

where $(x_0, v_0) = (x - \tau_r(x, -v)v, v)$ and $h_r(x_1/r, v_1) = \sin(n_{x_1}, v_1)$, and m is the probability measure on $\mathbf{K}$ obtained in Theorem 2.8.2.

If this holds, the iterates of the transfer map T_r are described by the Markov chain with transition probability $2P(2s, h|h')$. This leads to a kinetic equation on an extended phase space for the Boltzmann–Grad limit of the periodic Lorentz gas in space dimension 2:

$$F(t, x, v, s, h) =$$

  density of particles with velocity v and position x at time t

  that will hit an obstacle after time s, with impact parameter h.

Theorem 2.9.1 (Caglioti–Golse [10, 11]). *Assume (H), and let $f^{\mathrm{in}} \geq 0$ belong to $C_c(\mathbf{R}^2 \times \mathbf{S}^1)$. Then one has*

$$f_r \longrightarrow \int_0^\infty \int_{-1}^1 F(\,\cdot\,, \,\cdot\,, \,\cdot\,, s, h)\, ds\, dh \quad \text{in } L^\infty(\mathbf{R}_+ \times \mathbf{R}^2 \times \mathbf{S}^1) \ \text{weak-}*$$

in the limit as $r \to 0^+$, where $F \equiv F(t, x, v, s, h)$ is the solution of

$$(\partial_t + v \cdot \nabla_x - \partial_s)F(t, x, v, s, h)$$
$$= \int_{-1}^1 2P(2s, h|h')F(t, x, R[\pi - 2\arcsin(h')]v, 0, h')\, dh',$$

$$F(0, x, v, s, h) = f^{\mathrm{in}}(x, v) \int_{2s}^\infty \int_{-1}^1 P(\tau, h|h')\, dh'\, d\tau,$$

with (x, v, s, h) running through $\mathbf{R}^2 \times \mathbf{S}^1 \times \mathbf{R}_+^* \times (-1, 1)$. The notation $R[\theta]$ designates the rotation of an angle θ.

Let us briefly sketch the computation leading to the kinetic equation above in the extended phase space $\mathcal{Z} = \mathbf{R}^2 \times \mathbf{S}^1 \times \mathbf{R}_+ \times [-1, 1]$.

In the limit as $r \to 0^+$, the sequence $(b_r^n(x, v))_{n \geq 1}$ converges to a sequence of i.i.d. random variables with values in $\mathbf{K} = [0, 1] \times \{\pm 1\}$, according to assumption (H).

Then, for each $s_0 > 0$ and $h_0 \in [-1, 1]$, we construct a Markov chain $(s_n, h_n)_{n \geq 1}$ with values in $\mathbf{R}_+ \times [-1, 1]$ in the following manner:

$$(s_n, h_n) = \mathbf{T}_{b_n}(h_{n-1}), \quad n \geq 1.$$

Now we define the jump process (X_t, V_t, S_t, H_t) starting from (x, v, s, h) in the following manner. First pick a trajectory of the sequence $\mathbf{b} = (b_n)_{n \geq 1}$; then, for each $s > 0$ and each $h \in [-1, 1]$, set

$$(s_0, h_0) = (s, h).$$

Define then inductively s_n and h_n for $n \geq 1$ by the formula above, together with

$$\sigma_n = s_0 + \cdots + s_{n-1}, \quad n \geq 1,$$

and

$$v_n = R[2\arcsin(h_{n-1}) - \pi]v_{n-1}, \quad n \geq 1.$$

With the sequence $(v_n, s_n, h_n)_{n \geq 1}$ so defined, we next introduce the formulas for (X_t, V_t, S_t, H_t):

- While $0 \leq t < \tau$, we set

$$X_t(x, v, s, h) = x + t\omega, \qquad S_t(x, v, s, h) = s - t,$$
$$V_t(x, v, s, h) = v, \qquad H_t(x, v, s, h) = h.$$

- For $\sigma_n < t < \sigma_{n+1}$, we set

$$X_t(x, v, s, h) = x + (t - \sigma_n)v_n,$$
$$V_t(x, v, s, h) = v_n,$$
$$S_t(x, v, s, h) = \sigma_{n+1} - t,$$
$$H_t(x, v, s, h) = h_n.$$

To summarize, the prescription above defines, for each $t \geq 0$, a map denoted T_t:

$$\mathcal{Z} \times \mathbf{K}^{\mathbf{N}^*} \ni (x, v, s, h, \mathbf{b}) \longmapsto T_t(x, \omega, \tau, h) = (X_t, V_t, S_t, H_t) \in Z$$

that is piecewise continuous in $t \in \mathbf{R}_+$.

Denote by $F^{\mathrm{in}} \equiv F^{\mathrm{in}}(x, v, s, h)$ the initial distribution function in the extended phase space $\mathcal{Z}$, and by $\chi \equiv \chi(x, v, s, h)$ an observable —without loss of generality, we assume that $\chi \in C_c^\infty(Z)$.

Define $F(t, \cdot, \cdot, \cdot, \cdot)$ by the formula

$$\iiiint_Z \chi(x, v, s, h) F(t, dx, dv, ds, dh)$$
$$= \iiiint_Z \mathbf{E}[\chi(T_t(x, v, s, h))] F^{\mathrm{in}}(x, v, s, h)\, dx\, dv\, ds\, dh,$$

where $\mathbf{E}$ designates the expectation on trajectories of the sequence of i.i.d. random variables $\mathbf{b} = (b_n)_{n>1}$.

In other words, $F(t, \cdot, \cdot, \cdot, \cdot)$ is the image under the map T_t of the measure $\mathrm{Prob}(d\mathbf{b}) F^{\mathrm{in}}(x, v, s, h)$, where

$$\mathrm{Prob}(d\mathbf{b}) = \prod_{n \geq 1} dm(b_n).$$

Set $g(t, x, v, s, h) = \mathbf{E}[\chi(T_t(x, v, s, h))]$; one has

$$g(t, x, v, s, h) = \mathbf{E}[\mathbf{1}_{t<s}\, \chi(T_t(x, v, s, h))] + \mathbf{E}[\mathbf{1}_{s<t}\, \chi(T_t(x, v, s, h))].$$

If $s > t$, there is no collision in the time interval $[0, t]$ for the trajectory considered, meaning that

$$T_t(x, v, s, h) = (x + tv, v, s - t, h).$$

Hence

$$\mathbf{E}[\mathbf{1}_{t<s}\, \chi(T_t(x, v, s, h))] = \chi(x + tv, v, s - t, h)\, \mathbf{1}_{t<s}.$$

On the other hand,

$$\mathbf{E}[\mathbf{1}_{s<t}\, \chi(T_t(x, v, s, h))] = \mathbf{E}[\mathbf{1}_{s<t}\, \chi(T_{(t-s)-0}T_{s+0}(x, v, s, h))]$$
$$= \mathbf{E}[\mathbf{1}_{s<t}\, \chi(T_{(t-s)-0}(x + sv, \mathcal{R}[\Delta(h)]v, s_1, h_1))]$$

with $(s_1, h_1) = \mathbf{T}_{b_1}(h)$ and $\Delta(h) = 2\arcsin(h) - \pi$.

Conditioning with respect to (s_1, h_1) shows that

$$\mathbf{E}[\mathbf{1}_{s<t}\, \chi(T_t(x, v, s, h))]$$
$$= \mathbf{E}[\mathbf{1}_{s<t}\, \mathbf{E}[\chi(T_{(t-s)-0}(x + sv, \mathcal{R}[\Delta(h)]v, s_1, h_1))|s_1, h_1]],$$

and

$$\mathbf{E}[\chi(T_{(t-s)-0}(x + sv, \mathcal{R}[\Delta(h)]v, s_1, h_1))|s_1, h_1]$$
$$= g(t - s, x + sv, \mathcal{R}[\Delta(h)]v, s_1, h_1).$$

Then

$$\mathbf{E}[\mathbf{1}_{s<t}\, \mathbf{E}[\chi(T_{(t-s)-0}(x + sv, \mathcal{R}[\Delta(h)]v, s_1, h_1))|s_1, h_1]]$$
$$= \mathbf{1}_{s<t} \int g(t - s, x + sv, \mathcal{R}[\Delta(h)]v, \mathbf{T}_{b_1}(h))]\, dm(b_1)$$
$$= \mathbf{1}_{s<t} \int g(t - s, x + sv, \mathcal{R}[\Delta(h)]v, s_1, h_1)] 2P(2s_1, h_1|h)\, ds_1\, dh_1.$$

Finally,

$$g(t, x, v, s, h) = \chi(x + tv, v, s - t, h)\, \mathbf{1}_{t<s}$$

$$+ \mathbf{1}_{s<t} \int g(t - s, x + sv, \mathcal{R}[\Delta(h)]v, s_1, h_1)]P(s_1, h_1|h)\, ds_1\, dh_1.$$

This formula represents the solution of the problem

$$(\partial_t - v \cdot \nabla_x + \partial_s)g = 0, \quad t, s > 0, \ x \in \mathbf{R}^2, \ s \in \mathbf{S}^1, \ |h| < 1,$$

$$g(t, x, s, 0, h) = \iint_{\mathbf{R}_+^* \times (-1,1)} P(s_1, h_1|h)g(t, x, v, s_1, h_1)\, ds_1\, dh_1,$$

$$g\big|_{t=0} = \chi.$$

The boundary condition for $s = 0$ can be replaced with a source term that is proportional to the Dirac measure $\delta_{s=0}$:

$$(\partial_t - v \cdot \nabla_x + \partial_s)g = \delta_{s=0} \iint_{\mathbf{R}_+^* \times (-1,1)} P(s_1, h_1|h)g(t, x, v, s_1, h_1)\, ds_1\, dh_1,$$

$$g\big|_{t=0} = \chi.$$

One concludes by observing that this problem is precisely the adjoint of the Cauchy problem in the theorem.

Let us conclude this section with a few bibliographical remarks. Although the Boltzmann–Grad limit of the periodic Lorentz gas is a fairly natural problem, it remained open for quite a long time after the pioneering work of G. Gallavotti on the case of a Poisson distribution of obstacles [18, 19].

Perhaps the main conceptual difficulty was to realize that this limit must involve a phase-space other than the usual phase-space of kinetic theory, i.e., the set $\mathbf{R}^2 \times \mathbf{S}^1$ of particle positions and velocities, and to find the appropriate extended phase-space where the Boltzmann–Grad limit of the periodic Lorentz gas can be described by an autonomous equation.

Already Theorem 5.1 in [9] suggested that, even in the simplest problem where the obstacles are absorbing —i.e., holes where particles disappear forever— the limit of the particle number density in the Boltzmann–Grad scaling cannot be described by an autonomous equation in the usual phase space $\mathbf{R}^2 \times \mathbf{S}^1$.

The extended phase space $\mathbf{R}^2 \times \mathbf{S}^1 \times \mathbf{R}_+ \times [-1, 1]$ and the structure of the limit equation were proposed for the first time by E. Caglioti and the author in 2006, and presented in several conferences —see for instance [23]; the first computation of the transition probability $P(s, h|h')$ (Theorem 2.8.1), together with the limit equation (Theorem 2.9.1) appeared in [10] for the first time. However, the theorem concerning the limit equation in [10] remained incomplete, as it was based on the independence assumption (H).

Shortly after that, J. Marklof and A. Strömbergsson proposed a complete derivation of the limit equation of Theorem 2.9.1 in a recent preprint [32]. Their analysis establishes the validity of this equation in any space dimension, using in particular the existence of a transition probability as in Theorem 2.8.1 in any space dimension, a result that they had proved in an earlier paper [31]. The method of proof in this article [31] avoided using continued or Farey fractions, and was based on group actions on lattices in the Euclidean space, and on an important theorem by M. Ratner implying some equidistribution results in homogeneous spaces. However, explicit computations (as in Theorem 2.8.1) of the transition probability in space dimension higher than 2 seem beyond reach at the time of this writing —see however [33] for computations of the 2-dimensional transition probability for more general interactions than hard sphere collisions.

Finally, the limit equation obtained in Theorem 2.9.1 is interesting in itself; some qualitative properties of this equation are discussed in [11].

Conclusion

Classical kinetic theory (Boltzmann theory for elastic, hard sphere collisions) is based on two fundamental principles:

a) Deflections in velocity at each collision are mutually independent and identically distributed.

b) Time intervals between collisions are mutually independent, independent of velocities, and exponentially distributed.

The Boltzmann–Grad limit of the periodic Lorentz gas provides an example of a non-classical kinetic theory where velocity deflections at each collision jointly form a Markov chain, and the time intervals between collisions are not independent of the velocity deflections.

In both cases, collisions are purely local and instantaneous events: indeed the Boltzmann–Grad scaling is such that the particle radius is negligible in the limit. The difference between these two cases is caused by the degree of correlation between obstacles, which is maximal in the second case since the obstacles are centered at the vertices of a lattice in the Euclidean space, whereas obstacles are assumed to be independent in the first case. It could be interesting to explore situations that are somehow intermediate between these two extreme cases —for instance, situations where long range correlations become negligible.

Otherwise, there remain several outstanding open problems related to the periodic Lorentz gas, such as

i) obtaining explicit expressions of the transition probability whose existence is proved by J. Marklof and A. Strömbergsson in [31], in all space dimensions, or

ii) treating the case where particles are accelerated by an external force —for instance the case of a constant magnetic field, so that the kinetic energy of particles remains constant.

Bibliography

[1] S. Blank, N. Krikorian, Thom's problem on irrational flows. Internat. J. Math. **4** (1993), 721–726.

[2] P. Bleher, Statistical properties of two-dimensional periodic Lorentz gas with infinite horizon. J. Statist. Phys. **66** (1992), 315–373.

[3] F. Boca, A. Zaharescu, The distribution of the free path lengths in the periodic two-dimensional Lorentz gas in the small-scatterer limit. Commun. Math. Phys. **269** (2007), 425–471.

[4] C. Boldrighini, L. A. Bunimovich, Ya. G. Sinai, On the Boltzmann equation for the Lorentz gas. J. Statist. Phys. **32** (1983), 477–501.

[5] L. Boltzmann, Weitere Studien über das Wärmegleichgewicht unter Gasmolekülen. Wiener Berichte **66** (1872), 275–370.

[6] J. Bourgain, F. Golse, B. Wennberg, On the distribution of free path lengths for the periodic Lorentz gas. Commun. Math. Phys. **190** (1998), 491-508.

[7] L. Bunimovich, Ya. G. Sinai, Statistical properties of Lorentz gas with periodic configuration of scatterers. Commun. Math. Phys. **78** (1980/81), 479–497.

[8] L. Bunimovich, N. Chernov, Ya. G. Sinai, Statistical properties of two-dimensional hyperbolic billiards. Russian Math. Surveys **46** (1991), 47–106.

[9] E. Caglioti, F. Golse, On the distribution of free path lengths for the periodic Lorentz gas III. Commun. Math. Phys. **236** (2003), 199–221.

[10] E. Caglioti, F. Golse, The Boltzmann–Grad limit of the periodic Lorentz gas in two space dimensions. C. R. Math. Acad. Sci. Paris **346** (2008), 477–482.

[11] E. Caglioti, F. Golse, On the Boltzmann-Grad Limit for the Two Dimensional Periodic Lorentz Gas. J. Stat. Phys. **141** (2010), 264–317.

[12] C. Cercignani, On the Boltzmann equation for rigid spheres. Transport Theory Statist. Phys. **2** (1972), 211–225.

[13] C. Cercignani, R. Illner, M. Pulvirenti, The mathematical theory of dilute gases. Applied Mathematical Sciences, 106. Springer-Verlag, New York, 1994.

[14] P. Dahlqvist, The Lyapunov exponent in the Sinai billiard in the small scatterer limit. Nonlinearity **10** (1997), 159–173.

[15] L. Desvillettes, V. Ricci, Nonmarkovianity of the Boltzmann–Grad limit of a system of random obstacles in a given force field. Bull. Sci. Math. **128** (2004), 39–46.

[16] P. Drude, Zur Elektronentheorie der Metalle. Annalen der Physik **306** (3) (1900), 566–613.

[17] H. S. Dumas, L. Dumas, F. Golse, Remarks on the notion of mean free path for a periodic array of spherical obstacles. J. Statist. Phys. **87** (1997), 943–950.

[18] G. Gallavotti, Divergences and approach to equilibrium in the Lorentz and the wind–tree–models. Phys. Rev. (2) **185** (1969), 308–322.

[19] G. Gallavotti, Rigorous theory of the Boltzmann equation in the Lorentz gas. Nota interna no. 358, Istituto di Fisica, Univ. di Roma (1972). Available as preprint mp-arc-93-304.

[20] G. Gallavotti, Statistical Mechanics: a Short Treatise, Springer, Berlin-Heidelberg (1999).

[21] F. Golse, On the statistics of free-path lengths for the periodic Lorentz gas. Proceedings of the XIVth International Congress on Mathematical Physics (Lisbon 2003), 439–446, World Scientific, Hackensack NJ, 2005.

[22] F. Golse, The periodic Lorentz gas in the Boltzmann–Grad limit. Proceedings of the International Congress of Mathematicians, Madrid 2006, vol. 3, 183–201, European Math. Soc., Zürich, 2006.

[23] F. Golse, The periodic Lorentz gas in the Boltzmann–Grad limit (joint work with J. Bourgain, E. Caglioti and B. Wennberg). Oberwolfach Report 54/2006, vol. 3 (2006), no. 4, 3214, European Math. Soc., Zürich, 2006.

[24] F. Golse, On the periodic Lorentz gas in the Boltzmann–Grad scaling. Ann. Faculté des Sci. Toulouse **17** (2008), 735–749.

[25] F. Golse, B. Wennberg, On the distribution of free path lengths for the periodic Lorentz gas II. M2AN Modél. Math. et Anal. Numér. **34** (2000), 1151–1163.

[26] H. Grad, Principles of the kinetic theory of gases, in: Handbuch der Physik, S. Flügge ed. Band XII, 205–294, Springer-Verlag, Berlin 1958.

[27] R. Illner, M. Pulvirenti, Global validity of the Boltzmann equation for two- and three-dimensional rare gas in vacuum. Erratum and improved result: "Global validity of the Boltzmann equation for a two-dimensional rare gas in vacuum" [Commun. Math. Phys. **105** (1986), 189–203] and "Global validity of the Boltzmann equation for a three-dimensional rare gas in vacuum" [ibid. **113** (1987), 79–85] by M. Pulvirenti. Commun. Math. Phys. **121** (1989), 143–146.

[28] A. Ya. Khinchin, Continued Fractions. The University of Chicago Press, Chicago, Ill.-London, 1964.

[29] O. E. Lanford III, Time evolution of large classical systems. In: Dynamical Systems, Theory and Applications (Rencontres, Battelle Res. Inst., Seattle, Wash., 1974). Lecture Notes in Phys., Vol. 38, Springer, Berlin, 1975, 1–111.

[30] H. Lorentz, Le mouvement des électrons dans les métaux. Arch. Néerl. **10** (1905), 336–371.

[31] J. Marklof, A. Strömbergsson, The distribution of free path lengths in the periodic Lorentz gas and related lattice point problems. Ann. of Math. **172** (2010), 1949–2033.

[32] J. Marklof, A. Strömbergsson, The Boltzmann–Grad limit of the periodic Lorentz gas. Ann. of Math. **174** (2011), 225–298.

[33] J. Marklof, A. Strömbergsson, Kinetic transport in the two-dimensional periodic Lorentz gas. Nonlinearity **21** (2008), 1413–1422.

[34] J. Clerk Maxwell, Illustration of the Dynamical Theory of Gases I & II. Philos. Magazine and J. of Science **19** (1860), 19–32 & **20** (1860) 21–37.

[35] J. Clerk Maxwell, On the Dynamical Theory of Gases. Philos. Trans. R. Soc. London, **157** (1867), 49–88.

[36] H. L. Montgomery, Ten Lectures on the Interface between Analytic Number Theory and Harmonic Analysis. CBMS Regional Conference Series in Mathematics, 84. American Mathematical Society, Providence, RI, 1994.

[37] V. Ricci, B. Wennberg, On the derivation of a linear Boltzmann equation from a periodic lattice gas. Stochastic Process. Appl. **111** (2004), 281–315.

[38] L. A. Santaló, Sobre la distribución probable de corpúsculos en un cuerpo, deducida de la distribución en sus secciones y problemas análogos. Revista Unión Mat. Argentina **9** (1943), 145–164.

[39] C. L. Siegel, Über Gitterpunkte in convexen Körpern und ein damit zusammenhängendes Extremalproblem. Acta Math. **65** (1935), 307–323.

[40] Ya. G. Sinai, Topics in Ergodic Theory. Princeton Mathematical Series, 44. Princeton University Press, Princeton, NJ, 1994.

[41] V. T. Sós, On the distribution mod. 1 of the sequence $n\alpha$. Ann. Univ. Sci. Univ. Budapest. Eötvös Math. **1** (1958), 127–134.

[42] J. Suranyi, Über die Anordnung der Vielfahren einer reelen Zahl mod. 1. Ann. Univ. Sci. Univ. Budapest. Eötvös Math. **1** (1958), 107–111.

[43] H. Spohn, The Lorentz process converges to a random flight process. Commun. Math. Phys. **60** (1978), 277–290.

[44] D. Szasz, Hard Ball Systems and the Lorentz Gas. Encyclopaedia of Mathematical Sciences, 101. Springer, 2000.

[45] S. Ukai, N. Point, H. Ghidouche, Sur la solution globale du problème mixte de l'équation de Boltzmann non linéaire. J. Math. Pures Appl. (9) **57** (1978), 203–229.

Chapter 3

The Boltzmann Equation in Bounded Domains

Yan Guo

3.1 Introduction

Boundary effects play a crucial role in the dynamics of gases governed by the Boltzmann equation:

$$\partial_t F + v \cdot \nabla_x F = Q(F, F), \tag{3.1.1}$$

where $F(t, x, v)$ is the distribution function for the gas particles at time $t \geq 0$, position $x \in \Omega$, and $v \in \mathbf{R}^3$. Throughout this chapter, the collision operator takes the form

$$Q(F_1, F_2) = \int_{\mathbf{R}^3} \int_{\mathbf{S}^2} |v - u|^\gamma F_1(u') F_2(v') q_0(\theta) \, d\omega \, du$$

$$- \int_{\mathbf{R}^3} \int_{\mathbf{S}^2} |v - u|^\gamma F_1(u) F_2(v) q_0(\theta) \, d\omega \, du$$

$$\equiv Q_{\text{gain}}(F_1, F_2) - Q_{\text{loss}}(F_1, F_2), \tag{3.1.2}$$

where $u' = u + [(v - u) \cdot \omega]\omega$, $v' = v - [(v - u) \cdot \omega]\omega$, $\cos\theta = (u - v) \cdot \omega / |u - v|$, $0 \leq \gamma \leq 1$ (hard potential) and $0 \leq q_0(\theta) \leq C|\cos\theta|$ (angular cutoff). The mathematical study of the particle-boundary interaction in a bounded domain and its effect on the global dynamics is one of the fundamental problems in the Boltzmann theory. There are four basic types of boundary conditions for $F(t, x, v)$ at the boundary $\partial\Omega$:

(1) Inflow injection, in which the incoming particles are prescribed.

(2) Bounce-back reflection, in which the particles bounce back with reversed velocity.

(3) Specular reflection, in which the particles bounce back specularly.

(4) Diffuse reflection (stochastic), in which the incoming particles are a probability average of the outgoing particles.

Due to its importance, there have been many contributions in the mathematical study of different aspects of the Boltzmann boundary value problems [1], [2], [3], [4], [6], [9], [10], [11], [15], [30], [31], [34], [37], [39], [46], [49], among others. See also the references in the books [8], [12] and [44].

According to Grad [28, p. 243], one of the basic problems in the Boltzmann study is to prove existence and uniqueness of its solutions, as well as their time-decay toward an absolute Maxwellian, in the presence of compatible physical boundary conditions in a general domain. In spite of those contributions to the study of Boltzmann boundary problems, there are fewer mathematical results of uniqueness, regularity, and time decay-rate for Boltzmann solutions toward a Maxwellian. In [41], it was announced that Boltzmann solutions near a Maxwellian would decay exponentially to it in a smooth bounded convex domain with specular reflection boundary conditions. Unfortunately, we are not aware of any complete proof for such a result [45]. In [30], global stability of the Maxwellian was established in a convex domain for diffusive boundary conditions. Recently, important progress has been made in [16] and [47] to establish an almost exponential decay rate for Boltzmann solutions with large amplitude for general collision kernels and general boundary conditions, provided that certain a-priori strong Sobolev estimates can be verified. Even though these estimates had been established for spatially periodic domains [22], [23] near Maxwellians, their validity is completely open for the Boltzmann solutions, even local in time, in a bounded domain. As a matter of fact, this kind of strong Sobolev estimates may not be expected for a general non-convex domain [23]. This is because even for simplest kinetic equations with the differential operator $v \cdot \nabla_x$, the phase boundary $\partial\Omega \times \mathbf{R}^3$ is always characteristic but not uniformly characteristic at the grazing set $\gamma_0 = \{(x, v) : x \in \partial\Omega \text{ and } v \cdot n(x) = 0\}$ where $n(x)$ is the outward normal at x. Hence it is very challenging and delicate to obtain regularity from the general theory of hyperbolic PDE. Moreover, in comparison with the half-space problems studied, for instance in [34], [49], the geometrical complication makes it difficult to employ spatial Fourier transforms in x. There are many cycles (bouncing characteristics) interacting with the boundary repeatedly, and analysis of such cycles is one of the key mathematical difficulties.

We aim to develop a unified $L^2 - L^\infty$ theory in the near Maxwellian regime, to establish exponential decay toward a normalized Maxwellian $\mu = e^{-\frac{1}{2}|v|^2}$, for all four basic types of boundary conditions in rather general domains. Consequently, uniqueness among these solutions can be obtained. For convex domains, these solutions are shown to be continuous away from the singular grazing set γ_0.

3.2 Domain and characteristics

We let $\Omega = \{x : \xi(x) < 0\}$ be connected and bounded with $\xi(x)$ a smooth function. We assume that $\nabla\xi(x) \neq 0$ at the boundary $\xi(x) = 0$. The outward normal vector at $\partial\Omega$ is given by

$$n(x) = \frac{\nabla\xi(x)}{|\nabla\xi(x)|}, \tag{3.2.1}$$

and it can be extended smoothly near $\partial\Omega = \{x : \xi(x) = 0\}$. We say that Ω is *real analytic* if ξ is real analytic in x. We define Ω to be *strictly convex* if there exists $c_\xi > 0$ such that

$$\partial_{ij}\xi(x)\zeta^i\zeta^j \geq c_\xi|\zeta|^2 \tag{3.2.2}$$

for all x such that $\xi(x) \leq 0$, and all $\zeta \in \mathbf{R}^3$. We say that Ω has a *rotational symmetry* if there are vectors x_0 and ϖ such that, for all $x \in \partial\Omega$,

$$\{(x - x_0) \times \varpi\} \cdot n(x) \equiv 0. \tag{3.2.3}$$

We denote the phase boundary in the space $\Omega \times \mathbf{R}^3$ by $\gamma = \partial\Omega \times \mathbf{R}^3$, and split it into the outgoing boundary γ_+, the incoming boundary γ_-, and the singular boundary γ_0 for grazing velocities:

$$\gamma_+ = \{(x, v) \in \partial\Omega \times \mathbf{R}^3 : n(x) \cdot v > 0\},$$
$$\gamma_- = \{(x, v) \in \partial\Omega \times \mathbf{R}^3 : n(x) \cdot v < 0\},$$
$$\gamma_0 = \{(x, v) \in \partial\Omega \times \mathbf{R}^3 : n(x) \cdot v = 0\}.$$

Given (t, x, v), let $[X(s), V(s)] = [X(s; t, x, v), V(s; t, x, v)] = [x + (s - t)v, v]$ be the trajectory (or the characteristics) for the Boltzmann equation (3.1.1):

$$\frac{dX(s)}{ds} = V(s), \qquad \frac{dV(s)}{ds} = 0, \tag{3.2.4}$$

with the initial condition $[X(t; t, x, v), V(t; t, x, v)] = [x, v]$.

Definition 1 (Backward exit time). For (x, v) with $x \in \bar{\Omega}$ such that there exists some $\tau > 0$ for which $x - sv \in \Omega$ for $0 \leq s \leq \tau$, we define $t_{\mathbf{b}}(x, v) > 0$ to be the last moment at which the back-time straight line $[X(s; 0, x, v), V(s; 0, x, v)]$ remains in the interior of Ω:

$$t_{\mathbf{b}}(x, v) = \sup\{\tau > 0 : x - sv \in \Omega \text{ for } 0 \leq s \leq \tau\}. \tag{3.2.5}$$

Clearly, for any $x \in \Omega$, $t_{\mathbf{b}}(x, v)$ is well-defined for all $v \in \mathbf{R}^3$. If $x \in \partial\Omega$, then $t_{\mathbf{b}}(x, v)$ is well-defined for all $v \cdot n(x) > 0$. For any (x, v), we use $t_{\mathbf{b}}(x, v)$ whenever it is well-defined. We have $x - t_{\mathbf{b}}v \in \partial\Omega$ and $\xi(x - t_{\mathbf{b}}v) = 0$. We also define

$$x_{\mathbf{b}}(x, v) = x(t_{\mathbf{b}}) = x - t_{\mathbf{b}}v \in \partial\Omega. \tag{3.2.6}$$

We always have $v \cdot n(x_{\mathbf{b}}) \leq 0$.

3.3   Boundary condition and conservation laws

In terms of the standard perturbation f such that $F = \mu + \sqrt{\mu}f$, the Boltzmann equation can be rewritten as

$$\{\partial_t + v \cdot \nabla + L\} f = \Gamma(f, f), \qquad f(0, x, v) = f_0(x, v),$$

where the standard linear Boltzmann operator (see [20]) is given by

$$Lf \equiv \nu f - Kf = -\frac{1}{\sqrt{\mu}}\{Q(\mu, \sqrt{\mu}f) + Q(\sqrt{\mu}f, \mu)\} = \nu f - \int \mathbf{k}(v, v')f(v')\, dv'$$

(3.3.1)

with the collision frequency $\nu(v) \equiv \int |v - u|^\gamma \mu(u) q_0(\theta)\, du\, d\theta \sim \{1 + |v|\}^\gamma$ for $0 \le \gamma \le 1$; and

$$\Gamma(f_1, f_2) = \frac{1}{\sqrt{\mu}}Q\left(\sqrt{\mu}f_1, \sqrt{\mu}f_2\right) \equiv \Gamma_{\text{gain}}(f_1, f_2) - \Gamma_{\text{loss}}(f_1, f_2).$$

(3.3.2)

In terms of f, we formulate the boundary conditions as follows.

(1) The inflow boundary condition: for $(x, v) \in \gamma_-$,

$$f|_{\gamma_-} = g(t, x, v).$$

(3.3.3)

(2) The bounce-back boundary condition: for $x \in \partial\Omega$,

$$f(t, x, v)|_{\gamma_-} = f(t, x, -v).$$

(3.3.4)

(3) Specular reflection: for $x \in \partial\Omega$, let

$$R(x)v = v - 2(n(x) \cdot v)n(x),$$

(3.3.5)

and

$$f(t, x, v)|_{\gamma_-} = f(x, v, v - 2(n(x) \cdot v)n(x)) = f(x, v, R(x)v).$$

(3.3.6)

(4) Diffuse reflection: assume the natural normalization with some constant $c_\mu > 0$,

$$c_\mu \int_{v \cdot n(x) > 0} \mu(v)|n(x) \cdot v|\, dv = 1.$$

(3.3.7)

Then, for $(x, v) \in \gamma_-$,

$$f(t, x, v)|_{\gamma_-} = c_\mu \sqrt{\mu(v)} \int_{v' \cdot n(x) > 0} f(t, x, v')\sqrt{\mu(v')}\{n(x) \cdot v'\}\, dv'.$$

(3.3.8)

For both the bounce-back and specular reflection conditions (3.3.4) and (3.3.6), it is well-known that both mass and energy are conserved for (3.1.1). Without loss of generality, we may always assume that the mass-energy conservation laws hold for $t \geq 0$, in terms of the perturbation f:

$$\int_{\Omega \times \mathbf{R}^3} f(t, x, v) \sqrt{\mu} \, dx \, dv = 0, \tag{3.3.9}$$

$$\int_{\Omega \times \mathbf{R}^3} |v|^2 f(t, x, v) \sqrt{\mu} \, dx \, dv = 0. \tag{3.3.10}$$

Moreover, if the domain Ω has *any* axis of rotation symmetry (3.2.3), then we further assume that the corresponding conservation of angular momentum is valid for all $t \geq 0$:

$$\int_{\Omega \times \mathbf{R}^3} \{(x - x_0) \times \varpi\} \cdot v f(t, x, v) \sqrt{\mu} \, dx \, dv = 0. \tag{3.3.11}$$

For the diffuse reflection (3.3.8), the mass conservation (3.3.9) is assumed to be valid.

3.4 Main results

We introduce the weight function for $\rho > 0$ and $\beta \in \mathbf{R}^1$,

$$w(v) = (1 + \rho^2 |v|^2)^\beta. \tag{3.4.1}$$

Theorem 2. *Assume that $w^{-2}\{1 + |v|\}^3 \in L^1$ in (3.4.1). There exists $\delta > 0$ such that, if $F_0 = \mu + \sqrt{\mu} f_0 \geq 0$ and*

$$\|w f_0\|_\infty + \sup_{0 \leq t \leq \infty} e^{\lambda_0 t} \|w g(t)\|_\infty \leq \delta$$

with $\lambda_0 > 0$, then there there exists a unique solution $F(t, x, v) = \mu + \sqrt{\mu} f \geq 0$ to the inflow boundary value problem (3.3.3) for the Boltzmann equation (3.1.1). There exists $0 < \lambda < \lambda_0$ such that

$$\sup_{0 \leq t \leq \infty} e^{\lambda t} \|w f(t)\|_\infty \leq C \left\{ \|w f_0\|_\infty + \sup_{0 \leq t \leq \infty} e^{\lambda_0 t} \|w g(t)\|_\infty \right\}.$$

Moreover, if Ω is strictly convex (3.2.2), and $f_0(x, v)$ is continuous except on γ_0, and $g(t, x, v)$ is continuous in $[0, \infty) \times \{\partial\Omega \times \mathbf{R}^3 \setminus \gamma_0\}$ with

$$f_0(x, v) = g(x, v) \text{ on } \gamma_-,$$

then $f(t, x, v)$ is continuous in $[0, \infty) \times \{\bar{\Omega} \times \mathbf{R}^3 \setminus \gamma_0\}$.

Theorem 3. *Assume that $w^{-2}\{1 + |v|\}^3 \in L^1$ in (3.4.1). Assume that the conservation of mass (3.3.9) and energy (3.3.10) are valid for f_0. Then there exists $\delta > 0$ such that if $F_0(x, v) = \mu + \sqrt{\mu} f_0(x, v) \geq 0$ and $||w f_0||_\infty \leq \delta$, there exists a unique solution $F(t, x, v) = \mu + \sqrt{\mu} f(t, x, v) \geq 0$ to the bounce-back boundary value problem (3.3.4) for the Boltzmann equation (3.1.1) such that*

$$\sup_{0 \leq t \leq \infty} e^{\lambda t} ||w f(t)||_\infty \leq C ||w f_0||_\infty$$

for some $\lambda > 0$. Moreover, if Ω is strictly convex (3.2.2), and initially $f_0(x, v)$ is continuous except on γ_0, and

$$f_0(x, v) = f_0(x, -v) \text{ on } \partial\Omega \times \mathbf{R}^3 \setminus \gamma_0,$$

then $f(t, x, v)$ is continuous in $[0, \infty) \times \{\bar\Omega \times \mathbf{R}^3 \setminus \gamma_0\}$.

Theorem 4. *Assume that $w^{-2}\{1 + |v|\}^3 \in L^1$ in (3.4.1). Assume that ξ is both strictly convex (3.2.2) and analytic, and the mass (3.3.9) and energy (3.3.10) are conserved for f_0. If Ω has any rotational symmetry (3.2.3), we require that the corresponding angular momentum (3.3.11) is conserved for f_0. Then there exists $\delta > 0$ such that if $F_0(x, v) = \mu + \sqrt{\mu} f_0(x, v) \geq 0$ and $||w f_0||_\infty \leq \delta$, there exists a unique solution $F(t, x, v) = \mu + \sqrt{\mu} f(t, x, v) \geq 0$ to the specular boundary value problem (3.3.6) for the Boltzmann equation (3.1.1) such that*

$$\sup_{0 \leq t \leq \infty} e^{\lambda t} ||w f(t)||_\infty \leq C ||w f_0||_\infty$$

for some $\lambda > 0$. Moreover, if $f_0(x, v)$ is continuous except on γ_0 and

$$f_0(x, v) = f_0(x, R(x)v) \text{ on } \partial\Omega,$$

then $f(t, x, v)$ is continuous in $[0, \infty) \times \{\bar\Omega \times \mathbf{R}^3 \setminus \gamma_0\}$.

Theorem 5. *Assume (3.3.7). Assume that $w^{-2}\{1 + |v|\}^3 \in L^1$ for the weight function w in (3.4.1). Assume the mass conservation (3.3.9) is valid for f_0. If $F_0(x, v) = \mu + \sqrt{\mu} f_0(x, v) \geq 0$ and $||w f_0||_\infty \leq \delta$ sufficiently small, then there exists a unique solution $F(t, x, v) = \mu + \sqrt{\mu} f(t, x, v) \geq 0$ to the diffuse boundary value problem (3.3.8) for the Boltzmann equation (3.1.1) such that*

$$\sup_{0 \leq t \leq \infty} e^{\lambda t} ||w f(t)||_\infty \leq C ||w f_0||_\infty$$

for some $\lambda > 0$. Moreover, if ξ is strictly convex, and if $f_0(x, v)$ is continuous except on γ_0 with

$$f_0(x, v)|_{\gamma_-} = c_\mu \sqrt{\mu} \int_{n_x \cdot v' > 0} f_0(x, v') \sqrt{\mu(v')} \{n(x) \cdot v'\} \, dv',$$

then $f(t, x, v)$ is continuous in $[0, \infty) \times \{\bar\Omega \times \mathbf{R}^3 \setminus \gamma_0\}$.

Our result extends the result in [30] to non-convex domains with a different, more direct approach.

3.5 Velocity lemma and analyticity

The Velocity Lemma plays the most important role in the study of continuity for cycles (bouncing generalized trajectories) in the specular case. It states that, in a strictly convex domain (3.2.2), the singular set γ_0 cannot be reached via the trajectory $dx/dt = v$, $dv/dt = 0$ from interior points inside Ω, and hence γ_0 does not really participate or interfere with the interior dynamics. No singularity would be created from γ_0 and it is possible to perform calculus for the back-time exit time $t_b(x, v)$. This is the foundation for future regularity study. Moreover, the Velocity Lemma also provides the lower bound away from the singular set γ_0, which leads to estimates for repeating bounces in the specular reflection case. Such a Velocity Lemma was first discovered in [22], [23], in the study of regularity of the Vlasov–Poisson (Maxwell) system with flat geometry. It was then generalized in [32] for the Vlasov–Poisson system in a ball, and it is the starting point for the construction of regular solutions to the Vlasov–Poisson system in a general convex domain [33] with specular boundary condition.

3.6 L^2 decay theory

Since no spatial Fourier transform is available, we first establish linear L^2 exponential decay estimates in Section 3.4 via a functional analytical approach. It turns out that it suffices to establish the following finite-time estimate:

$$\int_0^1 \|\mathbf{P}f(s)\|_\nu^2\, ds \le M \left\{ \int_0^1 \|\{\mathbf{I} - \mathbf{P}\}f(s)\|_\nu^2\, ds + \text{boundary contributions} \right\}$$
(3.6.1)

for any solution f to the linear Boltzmann equation

$$\partial_t f + v \cdot \nabla_x f + Lf = 0, \qquad f(0, x, v) = f_0(x, v)$$
(3.6.2)

with all four boundary conditions (3.3.3), (3.3.4), (3.3.6) and (3.3.8). Here, for any fixed (t, x), the standard projection $\mathbf{P}$ onto the hydrodynamic part is given by

$$\mathbf{P}f = \{a(t, x) + b(t, x) \cdot v + c(t, x)|v|^2\}\sqrt{\mu(v)},$$
(3.6.3)

$$\mathbf{P}_a f = a(t, x)\sqrt{\mu(v)}, \quad \mathbf{P}_b f = b(t, x)v\sqrt{\mu(v)}, \quad \mathbf{P}_c f = c(t, x)|v|^2\sqrt{\mu(v)},$$

and $\|\cdot\|_\nu$ is the weighted L^2 norm with the collision frequency $\nu(v)$.

Similar types of estimates like (3.6.1), but with strong Sobolev norms, have been established in recent years [20] via the so-called macroscopic equations for the coefficients a, b and c. The key of the analysis was based on the ellipticity for b which satisfies Poisson's equation

$$\Delta b = \partial^2 \{\mathbf{I} - \mathbf{P}\}f,$$
(3.6.4)

where ∂^2 denotes some second-order differential operator. In the presence of the boundary condition $b \cdot n(x) = 0$ (bounce-back and specular) or $b \equiv 0$ (inflow and diffuse) at $\partial\Omega$, such an ellipticity is very difficult to employ for the weak L^2 (instead of H^1) estimate for b in (3.6.4). This is due to lack of regularity of b in (3.6.4), and even the trace of b is hard to define. It remains an interesting open question if such a direct approach can work, which can lead to a more explicit estimate for the constant M in (3.6.1).

Instead, we employ the hyperbolic (transport) feature rather than elliptic feature of the problem to prove (3.6.1), by an indirect method of contradiction with an implicit constant M. We can find f_k such that, if (3.6.1) were not valid, then the normalized

$$Z_k(t, x, v) \equiv \frac{f_k(t, x, v)}{\sqrt{\int_0^1 \|\mathbf{P} f_k(s)\|_\nu^2 \, ds}}$$

satisfies $\int_0^1 \|\mathbf{P} Z_k(s)\|_\nu^2 \, ds \equiv 1$, and

$$\int_0^1 \|(\mathbf{I} - \mathbf{P}) Z_k(s)\|_\nu^2 \, ds \leq \frac{1}{k}. \tag{3.6.5}$$

Denote a weak limit of Z_k by Z. We expect that $Z = \mathbf{P} Z = 0$, by each of the four boundary conditions. The key is to prove that $Z_k \to Z$ strongly to reach a contradiction. By the averaging Lemma [18], we know that $Z_k(s) \to Z$ strongly in the interior of Ω. As expected, the most delicate part is to exclude possible concentration near the boundary $\partial\Omega$. Since Z_k is a solution to the transport equation, it then follows that, near $\partial\Omega$, the set of non-grazing velocity $v \cdot n(x) \neq 0$ can be reached via a trajectory from the interior of Ω, which implies that Z_k can be controlled on this non-grazing set with no concentration. On the other hand, over the remaining almost-grazing set $v \cdot n(x) \sim 0$, thanks to the fact (3.6.5), we know that

$$Z_k \sim \mathbf{P} Z_k = \{a_k(t, x) + b_k(t, x) \cdot v + c_k(t, x)|v|^2\} \sqrt{\mu(v)}.$$

We observe that such special form of velocity distribution $\mathbf{P} Z_k$ *cannot* have concentration on the almost-grazing set $v \cdot n(x) \sim 0$, and we therefore conclude (3.6.1). Clearly, the hyperbolic or the transport property is crucial to control boundary behaviors via the interior compactness of Z_k.

3.7 L^∞ decay theory

We study linear L^∞ decay for all four different types of boundary conditions: inflow, bounce-back, specular and diffuse (stochastic) reflection. In order to control the nonlinear term $\Gamma(f, f)$, we need to estimate the weighted L^∞ of wf. We recall that $L = \nu - K$, and study the L^∞ (pointwise) decay of the linear Boltzmann equation (3.6.2). We choose a weight function

$$h(t, x, v) = w(v) f(t, x, v), \tag{3.7.1}$$

and study the equivalent linear Boltzmann equation:

$$\{\partial_t + v \cdot \nabla_x + \nu - K_w\}h = 0, \qquad h(0, x, v) = h_0(x, v) \equiv w f_0, \qquad (3.7.2)$$

where

$$K_w h = w K \left(\frac{h}{w}\right), \qquad (3.7.3)$$

together with various boundary conditions (3.3.3), (3.3.4), (3.3.6), or (3.3.8). In bounce-back, specular, diffuse reflection, as well as in the inflow case with $g = 0$, we designate the semigroup $U(t)h_0$ to be the solution to (3.7.2), and the semigroup $G(t)h_0$ to be the solution to the simpler transport equation without collision K_w:

$$\{\partial_t + v \cdot \nabla_x + \nu\}h = 0, \qquad h(0, x, v) = h_0(x, v) = w f_0. \qquad (3.7.4)$$

Notice that neither $G(t)$ nor $U(t)$ is a strongly continuous semigroup in L^∞ [44].

We first obtain an explicit representation of $G(t)$ in the presence of various boundary conditions. Then we can obtain the explicit exponential decay estimate for $G(t)$. Moreover, we also establish the continuity for $G(t)$ with a forcing term q if Ω is strictly convex (3.2.2) based on the Velocity Lemma. To study the L^∞ decay for $U(t)$, we make use of the Duhamel Principle:

$$U(t) = G(t) + \int_0^t G(t - s_1) K_w U(s_1) \, ds_1. \qquad (3.7.5)$$

Following the pioneering work of Vidav [48] almost 40 years ago, we iterate (3.7.5) back to get:

$$U(t) = G(t) + \int_0^t G(t - s_1) K_w G(s_1) \, ds_1 + \int_0^t \int_0^{s_1} G(t - s_1) K_w G(s_1 - s) K_w U(s) \, ds \, ds_1, \qquad (3.7.6)$$

where certain compactness property was discovered for the last double integration. Such a compactness is a feature of the so-called 'A-smoothing operators' introduced in [48], which have many applications in the Boltzmann theory [26]. Moreover, such an 'A-smoothing' property provides an effective way to estimate the sharp growth rate of a wide class of semigroups, which is important in the study of nonlinear instability problems [27]. Recently, a similar iteration was employed and new compactness was observed in the so-called 'Mixture Lemma' for L^∞ decay of the Boltzmann equation, either for a whole or a half space [34], [35], [36]. Our idea here is to estimate the last double integral in terms of the L^2 norm of $f = h/w$, which decays exponentially by L^2 decay theory. The presence of different boundary conditions now leads to complicated bouncing trajectories. Each of the boundary conditions presents a different difficulty, as illustrated below.

For the inflow boundary condition (3.3.3), the back-time trajectory comes either from the initial plane or from the boundary. Even though, when $g \neq 0$, the solution operators for (3.7.4) and (3.7.2) are not semigroups, for any (t, x, v) a similar representation as $G(t - s_1)K_w G(s_1 - s)K_w U(s)$ is still possible. With the compactness property of K_w, the main contribution in (3.7.6) is roughly of the form

$$\int_0^t \int_0^{s_1} \int_{v',v'' \text{bounded}} |h(s, X(s; s_1, X(s_1; t, x, v), v'), v'')| \, dv' dv'' ds \, ds_1. \quad (3.7.7)$$

The v' integral is estimated by a change of variable introduced in [48],

$$y \equiv X(s; s_1, X(s_1; t, x, v), v') = x - (t - s_1)v - (s_1 - s)v'. \quad (3.7.8)$$

Since $\det(dy/dv') \neq 0$ is almost always true, the v'- and v''-integration in (3.7.7) can be bounded as follows, where $h = wf$:

$$\int_{\Omega, v'' \text{ bounded}} |h(s, y, v'')| \, dy \, dv'' \leq C \left(\int_{\Omega, v'' \text{ bounded}} |f(s, y, v'')|^2 \, dy \, dv'' \right)^{1/2}.$$

For bounce-back, specular or diffuse reflections, the characteristic trajectories repeatedly interact with the boundary. Instead of $X(s; t, x, v)$, we should use the generalized characteristics, defined as cycles, $X_{\mathbf{cl}}(s; t, x, v)$ in (3.7.7). The key question is, for any fixed (t, x, v), whether or not the change of variable

$$y \equiv X_{\mathbf{cl}}(s; s_1, X_{\mathbf{cl}}(s_1; t, x, v), v') \quad (3.7.9)$$

is valid, i.e., to determine if it is almost always true that

$$\det \left\{ \frac{dX_{\mathbf{cl}}(s; s_1, X_{\mathbf{cl}}(s_1; t, x, v), v')}{dv'} \right\} \neq 0. \quad (3.7.10)$$

The bounce-back cycles $X_{\mathbf{cl}}(s; t, x, v)$ from a given point (t, x, v) are relatively simple, just going back and forth between two fixed points $x_{\mathbf{b}}(x, v)$ and $x_{\mathbf{b}}(x_{\mathbf{b}}(x, v), -v)$. Now the change of variable (3.7.9) and (3.7.10) can be established by the study of the set $S_x(v)$.

The specular cycles $X_{\mathbf{cl}}(s; t, x, v)$ reflect repeatedly with the boundary, and $dX_{\mathbf{cl}}(s; s_1, X_{\mathbf{cl}}(s_1; t, x, v), v')/dv'$ is very complicated to compute and (3.7.10) is extremely difficult to verify, even in a convex domain. This is in part due to the fact that there is no apparent way to analyze $dX_{\mathbf{cl}}(s; s_1 X_{\mathbf{cl}}(s_1; t, x, v), v')/dv'$ inductively with finite bounces. To overcome such a difficulty, $\det(dv_k/dv_1)$ can be computed asymptotically in a delicate iterative fashion for special cycles almost tangential to the boundary, which undergo many small bounces near the boundary. It then follows that $\det\{dX_{\mathbf{cl}}(s; s_1, X_{\mathbf{cl}}(s_1; t, x, v), v')/dv'\} \neq 0$ for these special cycles. This crucial observation is then combined with analyticity of ξ to conclude that the set of $\det\{dX_{\mathbf{cl}}(s; s_1, X_{\mathbf{cl}}(s_1; x, v), v')/dv'\} = 0$ is arbitrarily small, and the change of variable (3.7.9) is almost always valid. Analyticity plays an important role in our proof, and it certainly is an interesting open question to remove such a restriction in the future.

The diffuse cycles $X_{\mathbf{cl}}(s;t,x,v)$ contain more and more independent variables and (3.7.7) involves their integrations. A change of variable similar to (3.7.8) is expected with respect to one of those independent variables. However, the main difficulty in this case is the L^∞ control of $G(t)$ which satisfies (3.7.4). The most natural L^∞ estimate for $G(t)$ is for the weight $w = \mu^{-\frac{1}{2}}$, in which the diffuse boundary condition takes the form

$$h(t,x,v) = c_\mu \int_{v' \cdot n(x) \geq 0} h(t,x,v')\mu(v')\{v' \cdot n(x)\}\, dv'$$

with $c_\mu \int_{v' \cdot n(x) > 0} \mu(v')\{v' \cdot n(x)\}\, dv' = 1$. However, such a weight makes the linear Boltzmann theory break down. For any (t,x,v), since there are always particles moving almost tangential to the boundary in the bounce-back reflection, it is impossible to reach down the initial plane no matter how many cycles the particles take. In other words, there is no explicit expression for $G(t)$ in terms of initial data completely. To establish the L^∞ estimate for the different weight $w = \{1+\rho^2|v|^2\}^\beta$ in (3.4.1), we make the crucial observation that the measure of those particles that cannot reach the initial plane after k bounces is small when k is large. We make use of the freedom parameter ρ in our weight function to control the L^∞ norm. Therefore we can obtain an approximate representation formula for $G(t)$ by the initial datum, with only a finite number of bounces.

Bibliography

[1] Arkeryd, L., On the strong L^1 trend to equilibrium for the Boltzmann equation, Studies in Appl. Math. **87** (1992), 283–288.

[2] Arkeryd, L., Cercignani, C., A global existence theorem for the initial boundary value problem for the Boltzmann equation when the boundaries are not isothermal, Arch. Rational Mech. Anal. **125** (1993), 271–288.

[3] Arkeryd, L., Esposito, R., Marra, R., and Nouri, A., Stability of the laminar solution of the Boltzmann equation for the Benard problem, Bull. Inst. Math. Academia Sinica **3** (2008), 51–97.

[4] Arkeryd, L., Esposito, R., Pulvirenti, M., The Boltzmann equation for weakly inhomogeneous data, Commun. Math. Phys. **111** (1987), no. 3, 393–407.

[5] Alexandre, R., Desvillettes, L., Villani, C., and Wennberg, B., Entropy dissipation and long-range interactions, Arch. Rational. Mech. Anal. **152** (2000), no. 4, 327–355.

[6] Arkeryd, L., Heintz, A., On the solvability and asymptotics of the Boltzmann equation in irregular domains, Comm. Partial Differential Equations **22** (1997), no. 11-12, 2129–2152.

[7] Beals, R., Protopopescu, V., Abstract time-depedendent transport equations, J. Math. Anal. Appl. **212** (1987), 370–405.

[8] Cercignani, C., The Boltzmann Equation and Its Applications, Springer-Verlag, 1988.

[9] Cercignani, C., Equilibrium states and the trend to equlibrium in a gas according to the Boltzmann equation, Rend. Mat. Appl. **10** (1990), 77–95.

[10] Cercignani, C., On the initial-boundary value problem for the Boltzmann equation, Arch. Rational Mech. Anal. **116** (1992), 307–315.

[11] Cannoe, R., Cercignani, C., A trace theorem in kinetic theory, Appl. Math. Letters **4** (1991), 63–67.

[12] Cercignani, C., Illner, R., and Pulvirenti, M., The Mathematical Theory of Dilute Gases, Springer-Verlag, 1994.

[13] Chernov, N., Markarian, R., Chaotic Billiards, AMS, 2006.

[14] Deimling, K., Nonlinear Functional Analysis, Springer-Verlag, 1988.

[15] Desvillettes, L., Convergence to equilibrium in large time for Boltzmann and BGK equations, Arch. Rational Mech. Anal. **110** (1990), 73–91.

[16] Desvillettes, L., Villani, C., On the trend to global equilibrium for spatially inhomogeneous kinetic systems: the Boltzmann equation, Invent. Math. **159** (2005), no. 2, 245–316.

[17] Diperna, R., Lions, P.-L., On the Cauchy problem for the Boltzmann equation, Ann. of Math. **130** (1989), 321–366.

[18] Diperna, R., Lions, P.-L., Global weak solution of Vlasov–Maxwell systems, Comm. Pure Appl. Math. **42** (1989), 729–757.

[19] Esposito, L., Guo, Y., Marra, R., Phase transition in a Vlasov–Boltzmann system, Commun. Math. Phys. **296** (2010), 1–33.

[20] Guo, Y., The Vlasov–Poisson–Boltzmann system near Maxwellians, Comm. Pure Appl. Math. **55** (2002) no. 9, 1104–1135.

[21] Guo, Y., The Vlasov–Maxwell–Boltzmann system near Maxwellians, Invent. Math. **153** (2003), no. 3, 593–630.

[22] Guo, Y., Singular solutions of the Vlasov–Maxwell system on a half line, Arch. Rational Mech. Anal. **131** (1995), no. 3, 241–304.

[23] Guo, Y., Regularity for the Vlasov equations in a half-space, Indiana Univ. Math. J. **43** (1994), no. 1, 255–320.

[24] Guo, Y., Jang, J., and Jiang, N., Acoustic limit of the Boltzmann equation in optimal scaling, Comm. Pure Appl. Math. **63** (2010), no. 3, 337–361.

[25] Glassey, R., The Cauchy Problems in Kinetic Theory, SIAM, 1996.

[26] Glassey, R., Strauss, W. A., Asymptotic stability of the relativistic Maxwellian, Publ. Res. Inst. Math. Sci. **29** (1993), no. 2, 301–347.

[27] Guo, Y., Strauss, W. A., Instability of periodic BGK equilibria, Comm. Pure Appl. Math. **48** (1995), no. 8, 861–894.

[28] Grad, H., Principles of the kinetic theory of gases, Handbuch der Physik XII (1958), 205–294.

[29] Grad, H., Asymptotic theory of the Boltzmann equation, II. Rarefied gas dynamics, 3rd Symposium, Paris, 1962, 26–59.

[30] Guiraud, J. P., An *H*-theorem for a gas of rigid spheres in a bounded domain. Théories cinétique classique et relativiste (1975), G. Pichon, ed., CNRS, Paris, 29–58.

[31] Hamdache, K., Initial boundary value problems for the Boltzmann equation. Global existence of weak solutions, Arch. Rational Mech. Anal. **119** (1992), 309–353.

[32] Hwang, H.-J., Regularity for the Vlasov–Poisson system in a convex domain, SIAM J. Math. Anal. **36** (2004), no. 1, 121–171.

[33] Hwang, H.-J., Velázquez, J., Global existence for the Vlasov–Poisson system in bounded domains, Arch. Rational Mech. Anal. **195** (2010), no. 3, 763–796.

[34] Liu, T.-P.; Yu, S.-H., Initial-boundary value problem for one-dimensional wave solutions of the Boltzmann equation, Comm. Pure Appl. Math. **60** (2007), no. 3, 295–356.

[35] Liu, T.-P., Yu, S.-H., Green's function of the Boltzmann equation, 3-D waves, Bull. Inst. Math. Acad. Sin. (N.S.) **1** (2006), no. 1, 1–78.

[36] Liu, T.-P., Yu, S.-H., The Green's function and large-time behavior of solutions for the one-dimensional Boltzmann equation, Comm. Pure Appl. Math. **57** (2004), no. 12, 1543–1608.

[37] Maslova, N. B., Nonlinear Evolution Equations, Kinetic Approach, World Scientific Publishing, Singapore, 1993.

[38] Mischler, S., On the initial boundary value problem for the Vlasov–Poisson–Boltzmann system, Commun. Math. Phys. **210** (2000) 447–466.

[39] Masmoudi, N., Saint-Raymond, L., From the Boltzmann equation to the Stokes–Fourier system in a bounded domain, Comm. Pure Appl. Math. **56** (2003), no. 9, 1263–1293.

[40] Shizuta, Y., On the classical solutions of the Boltzmann equation, Comm. Pure Appl. Math. **36** (1983), 705–754.

[41] Shizuta, Y., Asano, K., Global solutions of the Boltzmann equation in a bounded convex domain, Proc. Japan Acad. **53A** (1977), 3–5.

[42] Strain, R., Guo, Y., Exponential decay for soft potentials near Maxwellians, Arch. Rational Mech. Anal. **187** (2008), no. 2, 287–339.

[43] Tabachnikov, S., Billiards, SMF, 1995.

[44] Ukai, S., Solutions of the Boltzmann equation. Pattern and Waves – Qualitative Analysis of Nonlinear Differential Equations, Studies of Mathematics and its Applications vol. 18, Kinokuniya-North-Holland, Tokyo, 1986, 37–96.

[45] Ukai, S., private communications.

[46] Ukai, S., Asano, K., On the initial boundary value problem of the linearized Boltzmann equation in an exterior domain, Proc. Japan Acad. **56** (1980), 12–17.

[47] Villani, C., Hypocoercivity, Memoirs of the AMS, vol. 202, no. 950, 2009.

[48] Vidav, I., Spectra of perturbed semigroups with applications to transport theory, J. Math. Anal. Appl. **30** (1970), 264–279.

[49] Yang, T., Zhao, H.-J., A half-space problem for the Boltzmann equation with specular reflection boundary condition, Commun. Math. Phys. **255** (2005), no. 3, 683–726.

Chapter 4

The Concentration-Compactness Rigidity Method for Critical Dispersive and Wave Equations

Carlos E. Kenig[1]

4.1 Introduction

In these lectures I will describe a program (which I will call the concentration-compactness/rigidity method) that Frank Merle and I have been developing to study critical evolution problems. The issues studied center around global well-posedness and scattering. The method applies to nonlinear dispersive and wave equations in both defocusing and focusing cases. The method can be divided into two parts. The first part (the "concentration-compactness" part) is in some sense "universal" and works in similar ways for "all" critical problems. The second part (the "rigidity" part) has a "universal" formulation, but needs to be established individually for each problem. The method is inspired by the elliptic work on the Yamabe problem and by works of Merle, Martel–Merle and Merle–Raphäel in the nonlinear Schrödinger equation and generalized KdV equations.

To focus on the issues, let us first concentrate on the energy critical nonlinear Schrödinger equation (NLS) and the energy critical nonlinear wave equation (NLW). We thus have:

$$\begin{cases} i\,\partial_t u + \triangle u \pm |u|^{4/N-2}u = 0, & (x,t) \in \mathbb{R}^N \times \mathbb{R}, \\\\ u\big|_{t=0} = u_0 \in \dot{H}^1(\mathbb{R}^n), & N \geq 3, \end{cases} \tag{4.1.1}$$

[1]Supported in part by NSF grant DMS-0456583.

and

$$
\begin{cases}
\partial_t^2 u - \triangle u = \pm |u|^{4/N-2} u, & (x,t) \in \mathbb{R}^N \times \mathbb{R}, \\[2mm]
u\big|_{t=0} = u_0 \in \dot{H}^1(\mathbb{R}^n), \\[2mm]
\partial_t u\big|_{t=0} = u_1 \in L^2(\mathbb{R}^n), & N \geq 3.
\end{cases} \tag{4.1.2}
$$

In both cases, the "$-$" sign corresponds to the defocusing case, while the "$+$" sign corresponds to the focusing case. For (4.1.1), if u is a solution, so is $\frac{1}{\lambda^{N-2/2}} u\left(\frac{x}{\lambda}, \frac{t}{\lambda^2}\right)$. For (4.1.2), if u is a solution, so is $\frac{1}{\lambda^{N-2/2}} u\left(\frac{x}{\lambda}, \frac{t}{\lambda}\right)$. Both scalings leave invariant the energy spaces $\dot{H}^1$, $\dot{H}^1 \times L^2$ respectively, and that is why they are called energy critical. The energy which is conserved in this problem is

$$
E_\pm(u_0) = \frac{1}{2} \int |\nabla u_0|^2 \pm \frac{1}{2^*} \int |u_0|^{2^*}, \tag{NLS}
$$

$$
E_\pm((u_0, u_1)) = \frac{1}{2} \int |\nabla u_0|^2 + \frac{1}{2} \int |u_1|^2 \pm \frac{1}{2^*} \int |u_0|^{2^*}, \tag{NLW}
$$

where $\frac{1}{2^*} = \frac{1}{2} - \frac{1}{N} = \frac{N-2}{2N}$. The "$+$" corresponds to the defocusing case while the "$-$" corresponds to the focusing case.

In both problems, the theory of the local Cauchy problem has been understood for a while (in the case of (4.1.1), through the work of Cazenave–Weissler [7], while in the case of (4.1.2) through the works of Pecher [37], Ginibre–Velo [14], Ginibre–Velo–Soffer [13], and many others, for instance [3], [20], [34], [41], etc.). These works show that, say for (4.1.1), for any u_0 with $\|u_0\|_{\dot{H}^1} \leq \delta$, there exists a unique solution of (4.1.1) defined for all time and the solution scatters, i.e., there exist u_0^+, u_0^- in $\dot{H}^1$ such that

$$
\lim_{t \to \pm\infty} \left\| u(t) - e^{it\triangle} u_0^\pm \right\|_{\dot{H}^1} = 0.
$$

A corresponding result holds for (4.1.2). Moreover, for any initial data u_0 $((u_0, u_1))$ in the energy space, there exist $T_+(u_0)$, $T_-(u_0)$ such that there exists a unique solution in $(-T_-(u_0), T_+(u_0))$ and the interval is maximal (for (4.1.2), $(-T_-(u_0, u_1), T_+(u_0, u_1)))$. In both problems, there exists a crucial space-time norm (or "Strichartz norm"). For (4.1.1), on a time interval I, we define

$$
\|u\|_{S(I)} = \|u\|_{L_I^{2(N+2)/N-2} L_x^{2(N+2)/N-2}},
$$

while for (4.1.2) we have

$$
\|u\|_{S(I)} = \|u\|_{L_I^{2(N+1)/N-2} L_x^{2(N+1)/N-2}}.
$$

This norm is crucial, say for (4.1.1), because, if $T_+(u_0) < +\infty$, we must have

$$
\|u\|_{S((0, T_+(u_0)))} = +\infty;
$$

moreover, if $T_+(u_0) = +\infty$, u scatters at $+\infty$ if and only if $\|u\|_{S(0,+\infty)} < +\infty$. Similar results hold for (4.1.2). The question that attracted people's attention here is: What happens for large data? The question was first studied for (4.1.2) in the defocusing case, through works of Struwe [44] in the radial case, Grillakis [16], [17] in the general case, for the preservation of smoothness, and in the terms described here in the works of Shatah–Struwe [41], [42], Bahouri–Shatah [3], Bahouri–Gérard [2], Kapitansky [20], etc. The summary of these works is that (this was achieved in the early 1990s), for any pair $(u_0, u_1) \in \dot{H}^1 \times L^2$, in the defocusing case we have $T_\pm(u_0, u_1) = +\infty$ and the solution scatters. The corresponding results for (4.1.1) in the defocusing case took much longer. The first result was established by Bourgain [4] in 1998, who established the analogous result for u_0 radial, $N = 3, 4$, with Grillakis [18] showing preservation of smoothness for $N = 3$ and radial data. Tao extended these results to $N \geq 5$, u_0 radial [48]. Finally, Colliander–Keel–Staffilani–Takaoka–Tao proved this for $N = 3$ and all data u_0 [8], with extensions to $N = 4$ by Ryckman–Vişan [40] and to $N \geq 5$ by Vişan [54] in 2005.

In the focusing case, these results do not hold. In fact, for (4.1.2) H. Levine [33] showed in 1974 that in the focusing case, if $(u_0, u_1) \in \dot{H}^1 \times L^2$, $u_0 \in L^2$ and $E((u_0, u_1)) < 0$, there is always a break-down in finite time, i.e., $T_\pm(u_0, u_1) < \infty$. He showed this by an "obstruction" type of argument. Recently Krieger–Schlag–Tătaru [32] have constructed radial examples ($N = 3$), for which $T_\pm(u_0, u_1) < \infty$. For (4.1.1) a classical argument due to Zakharov and Glassey [15], based on the virial identity, shows the same result as H. Levine's if $\int |x|^2 |u_0|^2 < \infty$, $E(u_0) < 0$. Moreover, for both (4.1.1) and (4.1.2), in the focusing case we have the following static solution:

$$W(x) = \left(1 + \frac{|x|^2}{N(N-2)}\right)^{-(N-2)/2} \in \dot{H}^1(\mathbb{R}^N),$$

which solves the elliptic equation

$$\triangle W + |W|^{4/N-2} W = 0.$$

Thus, scattering need not occur for solutions that exist globally in time. The solution W plays an important role in the Yamabe problem (see [1] for instance) and it does so once more here. The results on which I am going to concentrate here are the following.

Theorem 1 (Kenig–Merle [23]). *For the focusing energy critical (NLS), $3 \leq N \leq 6$, consider $u_0 \in \dot{H}^1$ such that $E(u_0) < E(W)$, u_0 radial. Then:*

   i) *If $\|u_0\|_{\dot{H}^1} < \|W\|_{\dot{H}^1}$, the solution exists for all time and scatters.*

   ii) *If $\|u_0\|_{L^2} < \infty$, $\|u_0\|_{\dot{H}^1} > \|W\|_{\dot{H}^1}$, then $T_+(u_0) < +\infty$, $T_-(u_0) < +\infty$.*

Remark 1. Recently, Killip–Vişan [29] have combined the ideas of the proof of Theorem 2, as applied to NLS in [10], with another important new idea, to extend Theorem 1 to the non-radial case for $N \geq 5$.

The case where the radial assumption is not needed in dimensions $3 \leq N \leq 6$ is the one of (4.1.2). We have:

Theorem 2 (Kenig–Merle [24]). *For the focusing energy critical (NLW), where $3 \leq N \leq 6$, consider $(u_0, u_1) \in \dot{H}^1 \times L^2$ such that $E((u_0, u_1)) < E((W, 0))$. Then:*

i) *If $\|u_0\|_{\dot{H}^1} < \|W\|_{\dot{H}^1}$, the solution exists for all time and scatters.*

ii) *If $\|u_0\|_{\dot{H}^1} > \|W\|_{\dot{H}^1}$, then $T_\pm(u_0) < +\infty$.*

I will sketch the proofs of these two theorems and the outline of the general method in these lectures. The method has found other interesting applications:

Mass Critical NLS:

$$
\begin{cases}
i\,\partial_t u + \triangle u \pm |u|^{4/N} u = 0, & (x, t) \in \mathbb{R}^N \times \mathbb{R}, \\[2mm]
u\big|_{t=0} = u_0, & N \geq 3.
\end{cases}
\tag{4.1.3}
$$

Here, $\|u_0\|_{L^2}$ is the critical norm. The analog of Theorem 1 was obtained, for u_0 radial, by Tao–Vişan–Zhang [50], Killip–Tao–Vişan [28], Killip–Vişan-Zhang [30], using our proof scheme for $N \geq 2$. (In the focusing case one needs to assume $\|u_0\|_{L^2} < \|Q\|_{L^2}$, where Q is the ground state, i.e., the non-negative solution of the elliptic equation $\triangle Q + Q^{1+4/N} = Q$.) The case $N = 1$ is open.

Corotational wave maps into S^2, 4D Yang–Mills in the radial case: Consider the wave map system

$$
\Box u = A(u)(Du, Du) \perp T_u M
$$

where $u = (u^1, \dots, u^d) : \mathbb{R} \times \mathbb{R}^N \to M \hookrightarrow \mathbb{R}^d$, where the target manifold M is isometrically embedded in $\mathbb{R}^d$, and $A(u)$ is the second fundamental form for M at u. We consider the case $M = S^2 \subset \mathbb{R}^3$. The critical space here is $(u_0, u_1) \in \dot{H}^{N/2} \times \dot{H}^{N-2/2}$, so that when $N = 2$, the critical space is $\dot{H}^1 \times L^2$. It is known that for small data in $\dot{H}^1 \times L^2$ we have global existence and scattering (Tătaru [52], [53], Tao [47]). Moreover, Rodnianski–Sterbenz [39] and Krieger–Schlag–Tătaru [31] showed that there can be finite time blow-up for large data. In earlier work, Struwe [45] had considered the case of co-rotational maps. These are maps which have a special form. Writing the metric on S^2 in the form (ρ, θ), $\rho > 0$, $\theta \in S^1$, with $ds^2 = d\rho^2 + g(\rho)^2 d\theta^2$, where $g(\rho) = \sin \rho$, we consider, using (r, ϕ) as polar coordinates in $\mathbb{R}^2$, maps of the form $\rho = v(r, t)$, $\theta = \phi$. These are the co-rotational maps and Krieger–Schlag–Tătaru [31] exhibited blow-up for corotational maps. There is a stationary solution Q, which is a non-constant harmonic map of least energy. Struwe proved that if $E(v) \leq E(Q)$, then v and the corresponding wave map u are global in time. Using our method, in joint work of Cote–Kenig–Merle [9] we show that, in addition, there is an alternative: $v \equiv Q$ or the solution scatters. We also prove the corresponding results for radial solutions of the Yang–Mills equations in the critical energy space in $\mathbb{R}^4$ (see [9]).

Cubic NLS in 3D: Consider the classic cubic NLS in 3D,

$$\begin{cases} i\,\partial_t u + \triangle u \mp |u|^2 u = 0, & (x,t) \in \mathbb{R}^3 \times \mathbb{R}, \\[2mm] u\big|_{t=0} = u_0 \in \dot{H}^{1/2}(\mathbb{R}^3). \end{cases}$$

Here $\dot{H}^{1/2}$ is the critical space, "$-$" corresponds to defocusing and "$+$" to focusing. In the focusing case, Duyckaerts–Holmer–Roudenko [10] adapted our method to show that if $u_0 \in \dot{H}^1(\mathbb{R}^3)$ and $M(u_0)E(u_0) < M(Q)E(Q)$, where

$$M(u_0) = \int |u_0|^2, \qquad E(u_0) = \frac{1}{2}\int |\nabla u_0|^2 - \frac{1}{4}\int |u_0|^4,$$

and Q is the ground state, i.e., the positive solution to the elliptic equation

$$-Q + \triangle Q + |Q|^2 Q = 0,$$

then if $\|u_0\|_{L^2}\|\nabla u_0\|_{L^2} > \|Q\|_{L^2}\|\nabla Q\|_{L^2}$, we have "blow-up" in finite time, while if $\|u_0\|_{L^2}\|\nabla u_0\|_{L^2} < \|Q\|_{L^2}\|\nabla Q\|_{L^2}$, then u exists for all time and scatters. In joint work with Merle [25] we have considered the defocusing case. We have shown, using this circle of ideas, that if $\sup_{0<t<T_+(u_0)} \|u(t)\|_{\dot{H}^{1/2}} < \infty$, then $T_+(u_0) = +\infty$ and u scatters. We would like to point out that the fact that $T_+(u_0) = +\infty$ is analogous to the $L^{3,\infty}$ result of Escauriaza–Seregin–Sverak for Navier–Stokes [11].

4.2   The Schrödinger equation

We now turn to the proofs of Theorems 1 and 2. We start with Theorem 1. We are thus considering

$$\begin{cases} i\,\partial_t u + \triangle u + |u|^{4/N-2} u = 0, & (x,t) \in \mathbb{R}^N \times \mathbb{R}, \\[2mm] u\big|_{t=0} = u_0 \in \dot{H}^1. \end{cases} \qquad (4.2.1)$$

Let us start with a quick review of the "local Cauchy problem" theory. Besides the norm $\|f\|_{S(I)} = \|f\|_{L_I^{2(N+2)/N-2}L_x^{2(N+2)/N-2}}$ introduced earlier, we need the norm $\|f\|_{W(I)} = \|f\|_{L_I^{2(N+2)/N-2}L_x^{2(N+2)/N^2+4}}$.

Theorem 3 ([7], [23]). *Assume that $u_0 \in \dot{H}^1(\mathbb{R}^N)$, $\|u_0\|_{\dot{H}^1} \leq A$. Then, for $3 \leq N \leq 6$, there exists $\delta = \delta(A) > 0$ such that if $\big\|e^{it\triangle}u_0\big\|_{S(I)} \leq \delta$, $0 \in \overset{\circ}{I}$, there exists a unique solution to (4.2.1) in $\mathbb{R}^N \times I$, with $u \in C(I; \dot{H}^1)$ and $\|\nabla u\|_{W(I)} < +\infty$, $\|u\|_{S(I)} \leq 2\delta$. Moreover, the mapping $u_0 \in \dot{H}^1(\mathbb{R}^N) \to u \in C(I; \dot{H}^1)$ is Lipschitz.*

The proof is by fixed point. The key ingredients are the following "Strichartz estimates" [43], [21]:

$$\begin{cases} \left\| \nabla e^{it\Delta} u_0 \right\|_{W(-\infty,+\infty)} \leq C \left\| u_0 \right\|_{\dot{H}^1}, \\ \left\| \nabla \int_0^t e^{i(t-t')\Delta} g(\cdot,t')dt' \right\|_{W(-\infty,+\infty)} \leq C \left\| g \right\|_{L^2_t L^{2N/N+2}_x}, \\ \sup_t \left\| \nabla \int_0^t e^{i(t-t')\Delta} g(\cdot,t')dt' \right\|_{L^2} \leq C \left\| g \right\|_{L^2_t L^{2N/N+2}_x} \end{cases} \qquad (4.2.2)$$

and the following Sobolev embedding:

$$\| v \|_{S(I)} \leq C \| \nabla v \|_{W(I)}, \qquad (4.2.3)$$

and the observation that $\left| \nabla(|u|^{4/N-2}u) \right| \leq C |\nabla u| \, |u|^{4/N-2}$, so that

$$\left\| \nabla(|u|^{4/N-2}u) \right\|_{L^2_I L^{2N/N+2}_x} \lesssim \| u \|_{S(I)}^{4/N-2} \| \nabla u \|_{W(I)}.$$

Remark 2. Because of (4.2.2), (4.2.3), there exists $\tilde{\delta}$ such that if $\| u_0 \|_{\dot{H}^1} \leq \tilde{\delta}$, the hypothesis of the theorem is verified for $I = (-\infty,+\infty)$. Moreover, given $u_0 \in \dot{H}^1$, we can find I such that $\| e^{it\Delta} u_0 \|_{S(I)} < \delta$, so that the theorem applies. It is then easy to see that given $u_0 \in \dot{H}^1$, there exists a maximal interval $I = (-T_-(u_0), T_+(u_0))$ where $u \in C(I'; \dot{H}^1) \cap \{ \nabla u \in W(I') \}$ of all $I' \subset\subset I$ is defined. We call I the maximal interval of existence. It is easy to see that for all $t \in I$, we have

$$E(u(t)) = \frac{1}{2} \int |\nabla u(t)|^2 - \frac{1}{2^*} \int |u|^{2^*} = E(u_0).$$

We also have the "standard finite-time blow-up criterion": if $T_+(u_0) < \infty$, then $\| u \|_{S([0,T_+(u_0)))} = +\infty$.

We next turn to another fundamental result in the "local Cauchy theory", the so-called "Perturbation Theorem".

Perturbation Theorem 15 (see [49], [23], [22]). *Let $I = [0, L)$, $L \leq +\infty$, and $\tilde{u}$ defined on $\mathbb{R}^N \times I$ be such that*

$$\sup_{t \in I} \| \tilde{u} \|_{\dot{H}^1} \leq A, \qquad \| \tilde{u} \|_{S(I)} \leq M, \qquad \| \nabla \tilde{u} \|_{W(I)} < +\infty,$$

and verify (in the sense of the integral equation)

$$i\,\partial_t \tilde{u} + \triangle \tilde{u} + |\tilde{u}|^{4/N-2}\tilde{u} = e \quad on \ \mathbb{R}^N \times I,$$

and let $u_0 \in \dot{H}^1$ be such that $\| u_0 - \tilde{u}(0) \|_{\dot{H}^1} \leq A'$. Then there exists $\epsilon_0 = \epsilon_0(M, A, A')$ such that, if $0 \leq \epsilon \leq \epsilon_0$ and $\| \nabla e \|_{L^2_I L^{2N/N+2}_x} \leq \epsilon$, $\left\| e^{it\Delta}[u_0 - \tilde{u}(0)] \right\|_{S(I)} \leq \epsilon$, then there exists a unique solution u to (4.2.1) on $\mathbb{R}^N \times I$, such that

$$\| u \|_{S(I)} \leq C(A, A', M) \quad and \quad \sup_{t \in I} \| u(t) - \tilde{u}(t) \|_{\dot{H}^1} \leq C(A, A', M)(A' + \epsilon)^\beta,$$

where $\beta > 0$.

For the details of the proof, see [22]. This result has several important consequences:

Corollary 1. *Let $K \subset \dot{H}^1$ be such that $\overline{K}$ is compact. Then there exist $T_{+,\overline{K}}$, $T_{-,\overline{K}}$ such that for all $u_0 \in K$ we have $T_+(u_0) \geq T_{+,\overline{K}}$, $T_-(u_0) \geq T_{-,\overline{K}}$.*

Corollary 2. *Let $\tilde{u}_0 \in \dot{H}^1$, $\|\tilde{u}_0\|_{\dot{H}^1} \leq A$, and let $\tilde{u}$ be the solution of (4.2.1), with maximal interval $(-T_-(\tilde{u}_0), T_+(\tilde{u}_0))$. Assume that $u_{0,n} \to \tilde{u}_0$ in $\dot{H}^1$, with corresponding solution u_n. Then $T_+(\tilde{u}_0) \leq \underline{\lim} T_+(u_{0,n})$, $T_-(\tilde{u}_0) \leq \underline{\lim} T_-(u_{0,n})$ and for $t \in (-T_-(u_0), T_+(u_0))$, $u_n(t) \to u(t)$ in $\dot{H}^1$.*

Before we start with our sketch of the proof of Theorem 1, we will review the classic argument of Glassey [15] for blow-up in finite time. Thus, assume $u_0 \in \dot{H}^1$, $\int |x|^2 |u_0(x)|^2 \, dx < \infty$ and $E(u_0) < 0$. Let I be the maximal interval of existence. One easily shows that, for $t \in I$, $y(t) = \int |x|^2 |u(x,t)|^2 \, dx < +\infty$. In fact,

$$y'(t) = 4 \operatorname{Im} \int \overline{u} \nabla u \cdot x, \quad \text{and} \quad y''(t) = 8 \left[\int |\nabla u(x,t)|^2 - \int |u(x,t)|^{2^*} \right].$$

Hence, if $E(u_0) < 0$, $E(u(t)) = E(u_0) < 0$, so that

$$\frac{1}{2} \int |\nabla u(t)|^2 - |u(t)|^{2^*} = E(u_0) + \left(\frac{1}{2^*} - \frac{1}{2} \right) \int |u(t)|^{2^*} \leq E(u_0) < 0,$$

and $y''(t) < 0$. But then, if I is infinite, since $y(t) > 0$ we obtain a contradiction. We now start with our sketch of the proof of Theorem 1.

Step 1: Variational estimates. (These are not needed in defocusing problems.) Recall that $W(x) = (1 + |x|^2/N(N-2))^{-(N-2)/2}$ is a stationary solution of (4.2.1). It solves the elliptic equation $\triangle W + |W|^{4/N-2} W = 0$, $W \geq 0$, W is radially decreasing, $W \in \dot{H}^1$. By the invariances of the equation,

$$W_{\theta_0, x_0, \lambda_0}(x) = e^{i\theta_0} \lambda_0^{N-2/2} W(\lambda_0 (x - x_0))$$

is still a solution. Aubin and Talenti [1], [46] gave the following variational characterization of W: let C_N be the best constant in the Sobolev embedding $\|u\|_{L^{2^*}} \leq C_N \|\nabla u\|_{L^2}$. Then $\|u\|_{L^{2^*}} = C_N \|\nabla u\|_{L^2}$, $u \not\equiv 0$, if and only if $u = W_{\theta_0, x_0, \lambda_0}$ for some $(\theta_0, x_0, \lambda_0)$. Note that by the elliptic equation, $\int |\nabla W|^2 = \int |W|^{2^*}$. Also, $C_N \|\nabla W\| = \|W\|_{L^{2^*}}$, so that

$$C_N^2 \|\nabla W\|^2 = \left(\int |\nabla W|^2 \right)^{\frac{N-2}{N}}.$$

Hence, $\int |\nabla W|^2 = 1/C_N^N$, and

$$E(W) = \left(\frac{1}{2} - \frac{1}{2^*} \right) \int |\nabla W|^2 = \frac{1}{N C_N^N}.$$

Lemma 1. *Assume that* $||\nabla v|| < ||\nabla W||$ *and that* $E(v) \leq (1 - \delta_0)E(W)$, $\delta_0 > 0$.
Then there exists $\bar{\delta} = \bar{\delta}(\delta_0)$ *so that:*

  i) $||\nabla v||^2 \leq (1 - \bar{\delta})||\nabla W||^2$;

 ii) $\int |\nabla v|^2 - |v|^{2^*} \geq \bar{\delta}||\nabla v||^2$;

iii) $E(v) \geq 0$.

Proof. Let

$$f(y) = \frac{1}{2}y - \frac{C_N^{2^*}}{2^*}y^{2^*/2}, \quad \bar{y} = ||\nabla v||^2.$$

Note that $f(0) = 0$, $f(y) > 0$ for y near 0, $y > 0$, and that

$$f'(y) = \frac{1}{2} - \frac{C_N^{2^*}}{2^*}y^{2^*/2 - 1},$$

so that $f'(y) = 0$ if and only if $y = y_c = 1/C_N = ||\nabla W||^2$. Also, $f(y_c) = 1/(NC_N) = E(W)$. Since $0 \leq \bar{y} < y_c$, $f(\bar{y}) \leq (1 - \delta_0)f(y_c)$, f is non-negative and strictly increasing between 0 and y_c, and $f''(y_c) \neq 0$, we have $0 \leq f(\bar{y})$, $\bar{y} \leq (1 - \bar{\delta})y_c = (1 - \bar{\delta})||\nabla W||^2$. This shows i).

For ii), note that

$$
\begin{aligned}
\int |\nabla v|^2 - |v|^{2^*} &\geq \int |\nabla v|^2 - C_N^{2^*}\left(\int |\nabla v|^2\right)^{2^*/2} \\
&= \int |\nabla v|^2 \left[1 - C_N^{2^*}\left(\int |\nabla v|^2\right)^{2/N-2}\right] \\
&\geq \int |\nabla v|^2 \left[1 - C_N^{2^*}(1 - \bar{\delta})^{2/N-2}\left(\int |\nabla W|^2\right)^{2/N-2}\right] \\
&= \int |\nabla v|^2 \left[1 - (1 - \bar{\delta})^{2/N-2}\right],
\end{aligned}
$$

which gives ii).

Note from this that if $||\nabla u_0|| < ||\nabla W||$, then $E(u_0) \geq 0$, i.e., iii) holds. □

This static lemma immediately has dynamic consequences.

Corollary 3 (Energy Trapping). *Let* u *be a solution of* (4.2.1) *with maximal interval* I, $||\nabla u_0|| < ||\nabla W||$, $E(u_0) < E(W)$. *Choose* $\delta_0 > 0$ *such that* $E(u_0) \leq (1 - \delta_0)E(W)$. *Then, for each* $t \in I$, *we have:*

  i) $||\nabla u(t)||^2 \leq (1 - \bar{\delta})||\nabla W||$, $\quad E(u(t)) \geq 0$;

 ii) $\int |\nabla u(t)|^2 - |u(t)|^{2^*} \geq \bar{\delta}\int |\nabla u(t)|^2$ *("coercivity");*

iii) $E(u(t)) \approx ||\nabla u(t)||^2 \approx ||\nabla u_0||^2$, *with comparability constants which depend on δ_0 ("uniform bound").*

Proof. The statements follow from continuity of the flow, conservation of energy and Lemma 1. □

Note that iii) gives uniform bounds on $||\nabla u(t)||$. However, this is a long way from giving Theorem 1.

Remark 3. Let $u_0 \in \dot{H}^1$, $E(u_0) < E(W)$, but $||\nabla u_0||^2 > ||\nabla W||^2$. If we choose δ_0 so that $E(u_0) \leq (1 - \delta_0)E(W)$, we can conclude, as in the proof of Lemma 1, that $\int |\nabla u(t)|^2 \geq (1 + \bar{\delta}) \int |\nabla W|^2$, $t \in I$. But then,

$$\int |\nabla u(t)|^2 - |u(t)|^{2^*} = 2^* E(u_0) - \frac{2}{N-2} \int |\nabla u|^2$$

$$\leq 2^* E(W) - \frac{2}{N-2} \frac{1}{C_N^N} - \frac{2\bar{\delta}}{N-2} \frac{1}{C_N^N}$$

$$= -\frac{2\bar{\delta}}{(N-2)C_N^N} < 0.$$

Hence, if $\int |x|^2 |u_0(x)|^2 \, dx < \infty$, Glassey's proof shows that I cannot be infinite. If u_0 is radial, $u_0 \in L^2$, using a "local virial identity" (which we will see momentarily) one can see that the same result holds.

Step 2: Concentration-compactness procedure. We now turn to the proof of i) in Theorem 1. By our variational estimates, if $E(u_0) < E(W)$, $||\nabla u_0||^2 < ||\nabla W||^2$, if δ_0 is chosen so that $E(u_0) \leq (1 - \delta_0)E(W)$, recall that

$$E(u(t)) \approx ||\nabla u(t)||^2 \approx ||\nabla u_0||^2,$$

$t \in I$, with constants depending only on δ_0. Recall also that if $||\nabla u_0||^2 < ||\nabla W||^2$, $E(u_0) \geq 0$. It now follows from the "local Cauchy theory" that if $||\nabla u_0||^2 < ||\nabla W||^2$ and $E(u_0) \leq \eta_0$, η_0 small, then $I = (-\infty, +\infty)$ and $||u||_{S(-\infty, +\infty)} < \infty$, so that u scatters. Consider now

$$G = \{E : 0 < E < E(W) :$$

$$\text{if } ||\nabla u_0||^2 < ||\nabla W||^2 \text{ and } E(u_0) < E, \text{ then } ||u||_{S(I)} < \infty\}$$

and $E_c = \sup G$. Then $0 < \eta_0 \leq E_c \leq E(W)$ and if $||\nabla u_0||^2 < ||\nabla W||^2$, $E(u_0) < E_c$, $I = (-\infty, +\infty)$, u scatters and E_c is optimal with this property. Theorem 1 i) is the statement $E_c = E(W)$. We now assume $E_c < E(W)$ and will reach a contradiction. We now develop the concentration-compactness argument:

Proposition 1. *There exists $u_{0,c} \in \dot{H}^1$, $||\nabla u_{0,c}||^2 < ||\nabla W||^2$, with $E(u_{0,c}) = E_c$, such that, for the corresponding solution u_c, we have $||u_c||_{S(I)} = +\infty$.*

Proposition 2. *For any u_c as in Proposition 1, with (say) $\|u_c\|_{S(I_+)} = +\infty$, $I_+ = I \cap [0, +\infty)$, there exist $x(t)$, $t \in I_+$, $\lambda(t) \in \mathbb{R}^+$, $t \in I_+$, such that*

$$K = \left\{ v(x,t) = \frac{1}{\lambda(t)^{N-2/2}} u\left(\frac{x - x(t)}{\lambda(t)}, t\right), \ t \in I_+ \right\}$$

has compact closure in $\dot{H}^1$.

The proof of Propositions 1 and 2 follows a "general procedure" which uses a "profile decomposition", the variational estimates and the "Perturbation Theorem". The idea of the decomposition is somehow a time-dependent version of the concentration-compactness method of P. L. Lions, when the "local Cauchy theory" is done in the critical space. It was introduced independently by Bahouri–Gérard [2] for the wave equation and by Merle–Vega for the L^2 critical NLS [35]. The version needed for Theorem 1 is due to Keraani [27]. This is the evolution analog of the elliptic "bubble decomposition", which goes back to work of Brézis–Coron [5].

Theorem 4 (Keraani [27]). *Let $\{v_{0,n}\} \subset \dot{H}^1$, with $\|v_{0,n}\|_{\dot{H}^1} \leq A$. Assume that $\|e^{it\Delta} v_{0,n}\|_{S(-\infty,+\infty)} \geq \delta > 0$. Then there exists a subsequence of $\{v_{0,n}\}$ and a sequence $\{V_{0,j}\}_{j=1}^\infty \subset \dot{H}^1$ and triples $\{(\lambda_{j,n}, x_{j,n}, t_{j,n})\} \subset \mathbb{R}^+ \times \mathbb{R}^N \times \mathbb{R}$, with*

$$\frac{\lambda_{j,n}}{\lambda_{j',n}} + \frac{\lambda_{j',n}}{\lambda_{j,n}} + \frac{|t_{j,n} - t_{j',n}|}{\lambda_{j,n}^2} + \frac{|x_{j,n} - x_{j',n}|}{\lambda_{j,n}} \xrightarrow[n \to \infty]{} \infty,$$

for $j \neq j'$ (we say that $\{(\lambda_{j,n}, x_{j,n}, t_{j,n})\}$ is orthogonal), such that

   i) $\|V_{0,1}\|_{\dot{H}^1} \geq \alpha_0(A) > 0$.

   ii) *If $V_j^l(x,t) = e^{it\Delta} V_{0,j}$, then we have, for each J,*

$$v_{0,n} = \sum_{j=1}^J \frac{1}{\lambda_{j,n}^{N-2/2}} V_j^l\left(\frac{x - x_{j,n}}{\lambda_{j,n}}, -\frac{t_{j,n}}{\lambda_{j,n}^2}\right) + w_n^J,$$

   where $\lim\limits_{n} \|e^{it\Delta} w_n^J\|_{S(-\infty,+\infty)} \xrightarrow[J \to \infty]{} 0$, and for each $J \geq 1$ we have

   iii) $\|\nabla v_{0,n}\|^2 = \sum\limits_{j=1}^J \|\nabla V_{0,j}\|^2 + \|\nabla w_n^J\|^2 + o(1)$ *as $n \to \infty$ and*

$$E(v_{0,n}) = \sum_{j=1}^J E\left(V_j^l\left(-\frac{t_{j,n}}{\lambda_{j,n}^2}\right)\right) + E(w_n^J) + o(1) \ \text{as } n \to \infty.$$

Further general remarks:

Remark 4. Because of the continuity of $u(t)$, $t \in I$, in $\dot{H}^1$, in Proposition 2 we can construct $\lambda(t)$, $x(t)$ continuous in $[0, T_+(u_0))$, with $\lambda(t) > 0$.

Remark 5. Because of scaling and the compactness of $\overline{K}$ above, if $T_+(u_{0,c}) < \infty$, one always has that $\lambda(t) \geq C_0(K)/(T_+(u_0, c) - t)^{\frac{1}{2}}$.

Remark 6. If $T_+(u_{0,c}) = +\infty$, we can always find another (possibly different) critical element v_c with a corresponding $\tilde{\lambda}$ so that $\tilde{\lambda}(t) \geq A_0 > 0$ for $t \in [0, T_+(v_{0,c}))$. (Again by compactness of $\overline{K}$.)

Remark 7. One can use the "profile decomposition" to also show that there exists a decreasing function $g\colon (0, E_c] \to [0, +\infty)$ so that if $||\nabla u_0||^2 < ||\nabla W||^2$ and $E(u_0) \leq E_c - \eta$, then $||u||_{S(-\infty,+\infty)} \leq y(\eta)$.

Remark 8. In the "profile decomposition", if all the $v_{0,n}$ are radial, the $V_{0,j}$ can be chosen radial and $x_{j,n} \equiv 0$. We can repeat our procedure restricted to radial data and conclude the analog of Propositions 1 and 2 with $x(t) \equiv 0$.

The final step in the proof is then:

Step 3: Rigidity Theorem.

Theorem 5 (Rigidity). *Let $u_0 \in \dot{H}^1$, $E(u_0) < E(W)$, $||\nabla u_0||^2 < ||\nabla W||^2$. Let u be the solution of (4.2.1), with maximal interval $I = (-T_-(u_0), T_+(u_0))$. Assume that there exists $\lambda(t) > 0$, defined for $t \in [0, T_+(u_0))$, such that*

$$K = \left\{ v(x,t) = \frac{1}{\lambda(t)^{N-2/2}} u\left(\frac{x}{\lambda(t)}, t \right), \; t \in [0, T_+(u_0)) \right\}$$

has compact closure in $\dot{H}^1$. Assume also that, if $T_+(u_0) < \infty$,

$$\lambda(t) \geq C_0(K)/(T_+(u_0, c) - t)^{\frac{1}{2}} \; and$$

if $T_+(u_0) = \infty$, that $\lambda(t) \geq A_0 > 0$ for $t \in [0, +\infty)$. Then $T_+(u_0) = +\infty$, $u_0 \equiv 0$.

To prove this, we split two cases:

Case 1: $T_+(u_0) < +\infty$. (So that $\lambda(t) \to +\infty$ as $t \to T_+(u_0)$.)

Fix ϕ radial, $\phi \in C_0^\infty$, $\phi \equiv 1$ on $|x| \leq 1$, supp $\phi \subset \{|x| < 2\}$. Set $\phi_R(x) = \phi(x/R)$ and define

$$y_R(t) = \int |u(x,t)|^2 \phi_R(x) \, dx.$$

Then $y_r'(t) = 2 \operatorname{Im} \int \overline{u} \nabla u \nabla \phi_R$, so that

$$|y_R'(t)| \leq C \left(\int |\nabla u|^2 \right)^{1/2} \left(\int \frac{|u|^2}{|x|^2} \right)^{1/2} \leq C ||\nabla W||^2,$$

by Hardy's inequality and our variational estimates. Note that C is independent of R. Next, we note that, for each $R > 0$,

$$\lim_{t \uparrow T_+(u_0)} \int_{|x|<R} |u(x,t)|^2 \, dx = 0.$$

In fact, $u(x,t) = \lambda(t)^{N-2/2}v(\lambda(t)x,t)$, so that

$$\int_{|x|<R}|u(x,t)|^2 dx = \lambda(t)^{-2}\int_{|y|<R\lambda(t)}|v(y,t)|^2\,dy$$

$$= \lambda(t)^{-2}\int_{|y|<\epsilon R\lambda(t)}|v(y,t)|^2\,dy$$

$$+ \lambda(t)^{-2}\int_{\epsilon R\lambda(t)\le|y|<R\lambda(t)}|v(y,t)|^2\,dy$$

$$= A + B.$$

$$A \le \lambda(t)^{-2}(\epsilon R\lambda(t))^2||v||_{L^{2^*}}^2 \le C\epsilon^2 R^2||\nabla W||^2,$$

which tends to 0 as ϵ tends to 0.

$$B \le \lambda(t)^{-2}(R\lambda(t))^2||v||_{L^{2^*}(|y|\ge\epsilon R\lambda(t))}^2 \xrightarrow[t\to T_+(u_0)]{} 0,$$

(since $\lambda(t) \uparrow +\infty$ as $t \to T_+(u_0)$) using the compactness of $\overline{K}$. But then $y_R(0) \le CT_+(u_0)||\nabla W||^2$, by the fundamental theorem of calculus. Thus, letting $R \to \infty$, we see that $u_0 \in L^2$, but then, using the conservation of the L^2 norm, we see that $||u_0||_{L^2} = ||u(T_+(u_0))||_{L^2} = 0$, so that $u_0 \equiv 0$.

Case 2: $T_+(u_0) = +\infty$. First note that the compactness of $\overline{K}$, together with $\lambda(t) \ge A_0 > 0$, gives that, given $\epsilon > 0$, there exists $R(\epsilon) > 0$ such that, for all $t \in [0,+\infty)$,

$$\int_{|x|>R(\epsilon)}|\nabla u|^2 + |u|^{2^*} + \frac{|u|^2}{|x|^2} \le \epsilon.$$

Pick $\delta_0 > 0$ so that $E(u_0) \le (1-\delta_0)E(W)$. Recall that, by our variational estimates, we have that $\int|\nabla u(t)|^2 - |u(t)|^{2^*} \ge C_{\delta_0}||\nabla u_0||_{L^2}^2$. If $||\nabla u_0||_{L^2} \ne 0$, using the smallness of tails, we see that, for $R > R_0$,

$$\int_{|x|<R}|\nabla u(t)|^2 - |u(t)|^{2^*} \ge C_{\delta_0}||\nabla u_0||_{L^2}^2.$$

Choose now $\psi \in C_0^\infty$ radial with $\psi(x) = |x|^2$ for $|x| \le 1$, supp $\psi \subset \{|x| \le 2\}$. Define

$$z_R(t) = \int|u(x,t)|^2 R^2\psi(x/R)\,dx.$$

Similar computations to Glassey's blow-up proof give

$$z_R'(t) = 2R\,\mathrm{Im}\int\overline{u}\nabla u\nabla\psi(x/R)$$

and

$$z_R''(t) = 4 \sum_{l,j} \operatorname{Re} \int \partial_{x_l} \partial_{x_j} \psi(x/R) \, \partial_{x_l} u \, \partial_{x_j} \overline{u}$$
$$- \frac{1}{R^2} \int \triangle^2 \psi(x/R) |u|^2 - \frac{4}{N} \int \triangle \psi(x/R) |u|^{2^*}.$$

Note that $|z_R'(t)| \leq C_{\delta_0} R^2 \|\nabla u_0\|^2$, by Cauchy–Schwartz, Hardy's inequality and our variational estimates. On the other hand,

$$z_R''(t) \geq \left[\int_{|x| < R} |\nabla u(t)|^2 - |u(t)|^{2^*} \right]$$
$$- C \left(\int_{R \leq |x| \leq 2R} |\nabla u(t)|^2 + \frac{|u|^2}{|x|^2} + |u(t)|^{2^*} \right)$$
$$\geq C \|\nabla u_0\|^2,$$

for R large. Integrating in t, we obtain $z_R'(t) - z_R'(0) \geq Ct \|\nabla u_0\|^2$, but

$$|z_R'(t) - z_R'(0)| \leq 2CR^2 \|\nabla u_0\|^2,$$

which is a contradiction for t large, proving Theorem 1 i).

Remark 9. In the defocusing case, the proof is easier since the variational estimates are not needed.

Remark 10. It is quite likely that for $N = 3$, examples similar to those by P. Raphäel [38] can be constructed, of radial data u_0 for which $T_+(u_0) < \infty$ and u blows up exactly on a sphere.

4.3 The wave equation

We now turn to Theorem 2. We thus consider

$$\begin{cases} \partial_t^2 u - \triangle u = |u|^{4/N-2} u, & (x,t) \in \mathbb{R}^N \times \mathbb{R}, \\[2mm] u\big|_{t=0} = u_0 \in \dot{H}^1(\mathbb{R}^n), & \\[2mm] \partial_t u\big|_{t=0} = u_1 \in L^2(\mathbb{R}^n), & N \geq 3. \end{cases} \tag{4.3.1}$$

Recall that $W(x) = \left(1 + |x|^2/N(N-2)\right)^{-(N-2)/2}$ is a static solution that does not scatter. The general scheme of the proof is similar to the one for Theorem 1. We start out with a brief review of the "local Cauchy problem". We first consider the associated linear problem,

$$
\begin{cases}
\partial_t^2 w - \triangle w = h, \\[2mm]
w\big|_{t=0} = w_0 \in \dot{H}^1(\mathbb{R}^n), \\[2mm]
\partial_t w\big|_{t=0} = w_1 \in L^2(\mathbb{R}^N).
\end{cases}
\tag{4.3.2}
$$

As is well known (see [42] for instance), the solution is given by

$$
\begin{aligned}
w(x,t) &= \cos\left(t\sqrt{-\triangle}\right) w_0 + (-\triangle)^{-1/2} \sin\left(t\sqrt{-\triangle}\right) w_1 \\
&\quad + \int_0^t (-\triangle)^{-1/2} \sin\left((t-s)\sqrt{-\triangle}\right) h(s)\, ds \\
&= S(t)((w_0, w_1)) + \int_0^t (-\triangle)^{-1/2} \sin\left((t-s)\sqrt{-\triangle}\right) h(s)\, ds.
\end{aligned}
$$

The following are the relevant Strichartz estimates: for an interval $I \subset \mathbb{R}$, let

$$
\|f\|_{S(I)} = \|f\|_{L_I^{2(N+1)/N-2} L_x^{2(N+1)/N-2}},
$$

$$
\|f\|_{W(I)} = \|f\|_{L_I^{2(N+1)/N-1} L_x^{2(N+1)/N-1}}.
$$

Then (see [14], [24])

$$
\begin{aligned}
\sup_t \|(w(t), \partial_t w(t))\|_{\dot{H}^1 \times L^2} &+ \left\|D^{1/2} w\right\|_{W(-\infty,+\infty)} \\
+ \left\|\partial_t D^{-1/2} w\right\|_{W(-\infty,+\infty)} &+ \|w\|_{S(-\infty,+\infty)} \\
+ \|w\|_{L_t^{(N+2)/N-2} L_x^{2(N+2)/N-2}} & \\
\leq C \Big\{ \|(w_0, w_1)\|_{\dot{H}^1 \times L^2} &+ \|w\|_{L_t^{2(N+1)/N+3} L_x^{2(N+1)/N+3}} \Big\}.
\end{aligned}
\tag{4.3.3}
$$

Because of the appearance of $D^{1/2}$ in these estimates, we also need to use the following version of the chain rule for fractional derivatives (see [26]).

Lemma 2. *Assume* $F \in C^2$, $F(0) = F'(0) = 0$, *and that for all* a, b *we have* $|F'(a+b)| \leq C\{|F'(a)| + |F'(b)|\}$ *and* $|F''(a+b)| \leq C\{|F''(a)| + |F''(b)|\}$. *Then, for* $0 < \alpha < 1$, $\frac{1}{p} = \frac{1}{p_1} + \frac{1}{p_2}$, $\frac{1}{p} = \frac{1}{r_1} + \frac{1}{r_2} + \frac{1}{r_3}$, *we have*

    i) $\|D^\alpha F(u)\|_{L^p} \leq C \|F'(u)\|_{L^{p_1}} \|D^\alpha u\|_{L^{p_2}}$,

    ii) $\|D^\alpha(F(u) - F(v))\|_{L^p} \leq C \left[\|F'(u)\|_{L^{p_1}} + \|F'(v)\|_{L^{p_1}} \right] \|D^\alpha(u-v)\|_{L^{p_2}}$

       $+ C \left[\|F''(u)\|_{L^{r_1}} + \|F''(v)\|_{L^{r_1}} \right] \left[\|D^\alpha u\|_{L^{r_2}} + \|D^\alpha v\|_{L^{r_2}} \right] \|u-v\|_{L^{r_3}}$.

    Using (4.3.3) and this lemma, one can now use the same argument as for (4.2.1) to obtain:

Theorem 6 ([14], [20], [24] and [41]). *Assume that*

$$(u_0, u_1) \in \dot{H}^1 \times L^2, \quad \|(u_0, u_1)\|_{\dot{H}^1 \times L^2} \leq A.$$

Then, for $3 \leq N \leq 6$, there exists $\delta = \delta(A) > 0$ such that if $\|S(t)(u_0, u_1)\|_{S(I)} \leq \delta$, $0 \in \overset{\circ}{I}$, there exists a unique solution to (4.3.1) in $\mathbb{R}^N \times I$, with $(u, \partial_t u) \in C(I; \dot{H}^1 \times L^2)$ and $\|D^{1/2} u\|_{W(I)} + \|\partial_t D^{-1/2} u\|_{W(I)} < \infty$, $\|u\|_{S(I)} \leq 2\delta$. Moreover, the mapping $(u_0, u_1) \in \dot{H}^1 \times L^2 \to (u, \partial_t u) \in C(I; \dot{H}^1 \times L^2)$ is Lipschitz.

Remark 11. Again, using (4.3.3), if $\|(u_0, u_1)\|_{\dot{H}^1 \times L^2} \leq \delta$, the hypothesis of the theorem is verified for $I = (-\infty, +\infty)$. Moreover, given $(u_0, u_1) \in \dot{H}^1 \times L^2$, we can find $\overset{\circ}{I} \ni 0$ so that the hypothesis is verified on I. One can then define a maximal interval of existence $I = (-T_-(u_0, u_1), T_+(u_0, u_1))$, similarly to the case of (4.2.1). We also have the "standard finite-time blow-up criterion": if $T_+(u_0, u_1) < \infty$, then $\|u\|_{S(0, T_+(u_0, u_1))} = +\infty$. Also, if $T_+(u_0, u_1) = +\infty$, u scatters at $+\infty$ (i.e., $\exists (u_0^+, u_1^+) \in \dot{H}^1 \times L^2$ such that $\|(u(t), \partial_t u(t)) - S(t)(u_0^+, u_1^+)\|_{\dot{H}^1 \times L^2} \xrightarrow[t \uparrow +\infty]{} 0$) if and only if $\|u\|_{S(0, +\infty)} < +\infty$. Moreover, for $t \in I$, we have

$$E((u_0, u_1)) = \frac{1}{2} \int |\nabla u_0|^2 + \frac{1}{2} \int u_1^2 - \frac{1}{2^*} \int |u_0|^{2^*} = E((u(t), \partial_t u(t))).$$

It turns out that for (4.3.1) there is another very important conserved quantity in the energy space, namely momentum. This is crucial for us to be able to treat non-radial data. This says that, for $t \in I$, $\int \nabla u(t) \cdot \partial_t u(t) = \int \nabla u_0 \cdot u_1$. Finally, the analog of the "Perturbation Theorem" also holds in this context (see [22]). All the corollaries of the Perturbation Theorem also hold.

Remark 12 (**Finite speed of propagation**). Recall that if $R(t)$ is the forward fundamental solution for the linear wave equation, the solution for (4.3.2) is given by (see [42])

$$w(t) = \partial_t R(t) * w_0 + R(t) * w_1 - \int_0^t R(t - s) * h(s) \, ds,$$

where $*$ stands for convolution in the x variable. The finite speed of propagation is the statement that supp $R(\cdot, t)$, supp $\partial_t R(\cdot, t) \subset \overline{B(0, t)}$. Thus, if supp $w_0 \subset {}^C B(x_0, a)$, supp $w_1 \subset {}^C B(x_0, a)$, supp $h \subset {}^C [\bigcup_{0 \leq t \leq a} B(x_0, a - t) \times \{t\}]$, then $w \equiv 0$ on $\bigcup_{0 \leq t \leq a} B(x_0, a - t) \times \{t\}$. This has important consequences for solutions of (4.3.1). If $(u_0, u_1) \equiv (u_0', u_1')$ on $B(x_0, a)$, then the corresponding solutions agree on $\bigcup_{0 \leq t \leq a} B(x_0, a - t) \times \{t\} \cap \mathbb{R}^N \times (I \cap I')$.

We now proceed with the proof of Theorem 2. As in the case of (4.2.1), the proof is broken up in three steps.

Step1: Variational estimates. Here these are immediate from the corresponding ones in (4.2.1). The summary is (we use the notation $\mathcal{E}(v) = \frac{1}{2} \int |\nabla v|^2 - \frac{1}{2^*} \int |v|^{2^*}$):

Lemma 3. *Let* $(u_0, u_1) \in \dot{H}^1 \times L^2$ *be such that* $E((u_0, u_1)) \le (1 - \delta_0)E((W, 0))$, $\|\nabla u_0\|^2 < \|\nabla W\|^2$. *Let* u *be the corresponding solution of* (4.3.1), *with maximal interval* I. *Then there exists* $\bar{\delta} = \bar{\delta}(\delta_0) > 0$ *such that, for* $t \in I$, *we have*

i) $\|\nabla u(t)\| \le (1 - \bar{\delta})\|\nabla W\|$.

ii) $\int |\nabla u(t)|^2 - |u(t)|^{2^*} \ge \bar{\delta} \int |\nabla u(t)|^2$.

iii) $\mathcal{E}(u(t)) \ge 0$ *(and here* $E((u, \partial_t u)) \ge 0$).

iv) $E((u, \partial_t u)) \approx \|(u(t), \partial_t u(t))\|_{\dot{H}^1 \times L^2}^2 \approx \|(u_0, u_1)\|_{\dot{H}^1 \times L^2}^2$, *with comparability constants depending only on* δ_0.

Remark 13. If $E((u_0, u_1)) \le (1 - \delta_0)E((W, 0))$, $\|\nabla u_0\|^2 > \|\nabla W\|^2$, then, for $t \in I$, $\|\nabla u(t)\|^2 \ge (1 + \bar{\delta})\|\nabla W\|^2$. This follows from the corresponding result for (4.2.1).

We now turn to the proof of ii) in Theorem 2. We will do it for the case when $\|u_0\|_{L^2} < \infty$. For the general case, see [24]. We know that, in the situation of ii), we have

$$\int |\nabla u(t)|^2 \ge (1 + \bar{\delta}) \int |\nabla W|^2, \quad t \in I,$$

$$E((W, 0)) \ge E((u(t), \partial_t u)) + \tilde{\delta}_0.$$

Thus,

$$\frac{1}{2^*} \int |u(t)|^{2^*} \ge \frac{1}{2} \int (\partial_t u(t))^2 + \frac{1}{2} \int |\nabla u(t)|^2 - E((W, 0)) + \tilde{\delta}_0,$$

so that

$$\int |u(t)|^{2^*} \ge \frac{N}{N - 2} \int (\partial_t u(t))^2 + \frac{N}{N - 2} \int |\nabla u(t)|^2 - 2^* E((W, 0)) + 2^* \tilde{\delta}_0.$$

Let $y(t) = \int |u(t)|^2$, so that $y'(t) = 2 \int u(t) \, \partial_t u(t)$. A simple calculation gives

$$y''(t) = 2 \int \left\{ (\partial_t u)^2 - |\nabla u(t)|^2 + |u(t)|^{2^*} \right\}.$$

Thus,

$$y''(t) \ge 2 \int (\partial_t u)^2 + \frac{2N}{N - 2} \int (\partial_t u)^2 - 2 \cdot 2^* E((W, 0))$$

$$+ \tilde{\tilde{\delta}}_0 + \frac{2N}{N - 2} \int |\nabla u(t)|^2 - 2 \int |\nabla u(t)|^2$$

$$= \frac{4(N - 1)}{N - 2} \int (\partial_t u)^2 + \frac{4}{N - 2} \int |\nabla u(t)|^2$$

$$- \frac{4}{N - 2} \int |\nabla W|^2 + \tilde{\tilde{\delta}}_0$$

$$\ge \frac{4(N - 1)}{N - 2} \int (\partial_t u)^2 + \tilde{\tilde{\delta}}_0.$$

If $I \cap [0, +\infty) = [0, +\infty)$, there exists $t_0 > 0$ so that $y'(t_0) > 0$, $y'(t) > 0$, $t > t_0$. For $t > t_0$ we have

$$y(t)y''(t) \geq \frac{4(N-1)}{N-2} \int (\partial_t u)^2 \int u^2 \geq \left(\frac{N-1}{N-2}\right) y'(t)^2,$$

so that

$$\frac{y''(t)}{y'(t)} \geq \left(\frac{N-1}{N-2}\right) \frac{y'(t)}{y(t)},$$

$$y'(t) \geq C_0 y(t)^{(N-1)/(N-2)}, \text{ for } t > t_0.$$

But, since $N - 1/N - 2 > 1$, this leads to finite-time blow-up, a contradiction. We next turn to the proof of i) in Theorem 2.

Step 2: Concentration-compactness procedure. Here we proceed initially in an identical manner as in the case of (4.2.1), replacing the "profile decomposition" of Keraani [27] with the corresponding one for the wave equation, due to Bahouri–Gérard [2]. Thus, arguing by contradiction, we find a number E_c, with $0 < \eta_0 \leq E_c < E((W,0))$ with the property that if $E((u_0, u_1)) < E_c$, $\|\nabla u_0\|^2 < \|\nabla W\|^2$, $\|u\|_{S(I)} < \infty$ and E_c is optimal with this property. We will see that this leads to a contradiction. As for (4.2.1), we have:

Proposition 3. *There exist*

$$(u_{0,c}, u_{1,c}) \in \dot{H}^1 \times L^2, \quad \|\nabla u_{0,c}\|^2 < \|\nabla W\|^2, \quad E((u_{0,c}, u_{1,c})) = E_c$$

and such that for the corresponding solution u_c on (4.3.1) we have $\|u_c\|_{S(I)} = +\infty$.

Proposition 4. *For any u_c as in Proposition 3, with (say) $\|u_c\|_{S(I_+)} = +\infty$, $I_+ = I \cap [0, +\infty)$, there exist $x(t) \in \mathbb{R}^N$, $\lambda(t) \in \mathbb{R}^+$, $t \in I_+$, such that*

$$K = \left\{ v(x,t) = \left(\frac{1}{\lambda(t)^{N-2/2}} u_c \left(\frac{x - x(t)}{\lambda(t)}, t \right), \frac{1}{\lambda(t)^{N/2}} \partial_t u_c \left(\frac{x - x(t)}{\lambda(t)}, t \right) \right) : t \in I_+ \right\}$$

has compact closure in $\dot{H}^1 \times L^2$.

Remark 14. As in the case of (4.2.1), in Proposition 4 we can construct $\lambda(t)$, $x(t)$ continuous in $[0, T_+((u_{0,c}, u_{1,c})))$. Moreover, by scaling and compactness of $\overline{K}$, if $T_+((u_{0,c}, u_{1,c})) < \infty$, we have $\lambda(t) \geq C_0(K)/(T_+((u_{0,c}, u_{1,c})) - t)$. Also, if $T_+((u_{0,c}, u_{1,c})) = +\infty$, we can always find another (possibly different) critical element v_c, with a corresponding $\tilde{\lambda}$ so that $\tilde{\lambda}(t) \geq A > 0$, for $t \in [0, T_+((v_{0,c}, v_{1,c})))$, using the compactness of $\overline{K}$. We can also find $g : (0, E_c] \to [0, +\infty)$ decreasing so that if $\|\nabla u_0\|^2 < \|\nabla W\|^2$ and $E((u_{0,c}, u_{1,c})) \leq E_c - \eta$, then $\|u\|_{S(-\infty, +\infty)} \leq g(\eta)$.

Up to here, we have used, in Step 2, only Step 1 and "general arguments". To proceed further we need to use specific features of (4.3.1) to establish further properties of critical elements.

The first one is a consequence of the finite speed of propagation and the compactness of $\overline{K}$.

Lemma 4. *Let u_c be a critical element as in Proposition 4, with $T_+((u_{0,c}, u_{1,c})) < +\infty$. (We can assume, by scaling, that $T_+((u_{0,c}, u_{1,c})) = 1$.) Then there exists $\overline{x} \in \mathbb{R}^N$ such that $\operatorname{supp} u_c(\,\cdot\,, t), \operatorname{supp} \partial_t u_c(\,\cdot\,, t) \subset B(\overline{x}, 1 - t), 0 < t < 1$.*

In order to prove this lemma, we will need the following consequence of the finite speed of propagation:

Remark 15. Let $(u_0, u_1) \in \dot{H}^1 \times L^2$, $\|(u_0, u_1)\|_{\dot{H}^1 \times L^2} \leq A$. If, for some $M > 0$ and $0 < \epsilon < \epsilon_0(A)$, we have

$$\int_{|x| \geq M} |\nabla u_0|^2 + |u_1|^2 + \frac{|u_0|^2}{|x|^2} \leq \epsilon,$$

then for $0 < t < T_+(u_0, u_1)$ we have

$$\int_{|x| \geq \frac{3}{2}M + t} |\nabla u(t)|^2 + |\partial_t u(t)|^2 + |u(t)|^{2^*} + \frac{|u(t)|^2}{|x|^2} \leq C\epsilon.$$

Indeed, choose $\psi_M \in C^\infty$, $\psi_M \equiv 1$ for $|x| \geq \frac{3}{2}M$, with $\psi_M \equiv 0$ for $|x| \leq M$. Let $u_{0,M} = u_0 \psi_M$, $u_{1,M} = u_1 \psi_M$. From our assumptions, we have $\|(u_{0,M}, u_{1,M})\|_{\dot{H}^1 \times L^2} \leq C\epsilon$. If $C\epsilon_0 < \tilde{\delta}$, where $\tilde{\delta}$ is as in the "local Cauchy theory", the corresponding solution u_M of (4.3.1) has maximal interval $(-\infty, +\infty)$ and $\sup_{t \in (-\infty, +\infty)} \|(u_M(t), \partial_t u_M(t))\|_{\dot{H}^1 \times L^2} \leq 2C\epsilon$. But, by finite speed of propagation, $u_M \equiv u$ for $|x| \geq \frac{3}{2}M + t$, $t \in [0, T_+(u_0, u_1))$, which proves the remark.

We turn to the proof of the lemma. Recall that $\lambda(t) \geq C_0(K)/(1 - t)$. We claim that, for any $R_0 > 0$,

$$\lim_{t \uparrow 1} \int_{|x + x(t)/\lambda(t)| \geq R_0} |\nabla u_c(x, t)|^2 + |\partial_t u_c(x, t)|^2 + \frac{|u_c(x, t)|^2}{|x|^2} = 0.$$

Indeed, if $\vec{v}(x, t) = \frac{1}{\lambda(t)^{N/2}} \left(\nabla u_c \left(\frac{x - x(t)}{\lambda(t)}, t \right), \partial_t u_c \left(\frac{x - x(t)}{\lambda(t)}, t \right) \right)$,

$$\int_{|x + x(t)/\lambda(t)| \geq R_0} |\nabla u_c(x, t)|^2 + |\partial_t u_c(x, t)|^2 = \int_{|y| \geq \lambda(t) R_0} |\vec{v}(x, t)|^2 \, dy \xrightarrow[t \uparrow 1]{} 0,$$

because of the compactness of $\overline{K}$ and the fact that $\lambda(t) \to +\infty$ as $t \to 1$. Because of this fact, using the remark backward in time, we have, for each $s \in [0, 1)$, $R_0 > 0$,

$$\lim_{t \uparrow 1} \int_{|x + x(t)/\lambda(t)| \geq \frac{3}{2} R_0 + (t - s)} |\nabla u_c(x, s)|^2 + |\partial_t u_c(x, s)|^2 = 0.$$

We next show that $|x(t)/\lambda(t)| \leq M$, $0 \leq t < 1$. If not, we can find $t_n \uparrow 1$ so that $|x(t_n)/\lambda(t_n)| \to +\infty$. Then, for $R > 0$, $\{|x| \leq R\} \subset \{|x + x(t_n)/\lambda(t_n)| \geq \frac{3}{2}R + t_n\}$ for n large, so that, passing to the limit in n, for $s = 0$, we obtain

$$\int_{|x| \leq R} |\nabla u_{0,c}|^2 + |u_{1,c}|^2 = 0,$$

a contradiction.

Finally, pick $t_n \uparrow 1$ so that $x(t_n)/\lambda(t_n) \to -\bar{x}$. Observe that, for every $\eta_0 > 0$, for n large enough, for all $s \in [0,1)$, $\{|x - \bar{x}| \geq 1 + \eta_0 - s\} \subset \{|x + x(t_n)/\lambda(t_n)| \geq \frac{3}{2}R_0 + (t_n - s)\}$, for some $R_0 = R_0(\eta_0) > 0$. From this we conclude that

$$\int_{|x-x_0|\geq 1+\eta_0-s} |\nabla u(x,s)|^2 + |\partial_s u(x,s)|^2 \, dx = 0,$$

which gives the claim.

Note that, after translation, we can assume that $\bar{x} = 0$. We next turn to a result which is fundamental for us to be able to treat non-radial data.

Theorem 7. *Let $(u_{0,c}, u_{1,c})$ be as in Proposition 4, with $\lambda(t)$, $x(t)$ continuous. Assume that either $T_+(u_{0,c}, u_{1,c}) < \infty$ or $T_+(u_{0,c}, u_{1,c}) = +\infty$, $\lambda(t) \geq A_0 > 0$. Then*

$$\int \nabla u_{0,c} \cdot u_{1,c} = 0.$$

In order to carry out the proof of this theorem, a further linear estimate is needed:

Lemma 5. *Let w solve the linear wave equation*

$$\begin{cases} \partial_t^2 w - \triangle w = h \in L_t^1 L_x^2(\mathbb{R}^{N+1}), \\[2mm] w\big|_{t=0} = w_0 \in \dot{H}^1(\mathbb{R}^n), \\[2mm] \partial_t w\big|_{t=0} = w_1 \in L^2(\mathbb{R}^N). \end{cases}$$

Then, for $|a| \leq 1/4$, we have

$$\sup_t \left\| \left(\nabla w \left(\frac{x_1 - at}{\sqrt{1-a^2}}, x', \frac{t - ax_1}{\sqrt{1-a^2}} \right), \partial_t w \left(\frac{x_1 - at}{\sqrt{1-a^2}}, x', \frac{t - ax_1}{\sqrt{1-a^2}} \right) \right) \right\|_{L^2(dx_1 dx')}$$

$$\leq C \left\{ \|w_0\|_{\dot{H}^1} + \|w_1\|_{L^2} + \|h\|_{L_t^1 L_x^2} \right\}.$$

The simple proof is omitted; see [24] for the details. Note that if u is a solution of (4.3.1), with maximal interval I and $I' \subset\subset I$, $u \in L_{I'}^{(N+2)/N-2} L_x^{2(N+2)/N-2}$, and since $\frac{4}{N-2} + 1 = \frac{N+2}{N-2}$, $|u|^{4/N-2} u \in L_t^1 L_x^2$. Thus, the conclusion of the lemma applies, provided the integration is restricted to $\left(\frac{x_1-at}{\sqrt{1-a^2}}, x', \frac{t-ax_1}{\sqrt{1-a^2}} \right) \in \mathbb{R}^N \times I'$.

Sketch of proof of Theorem 7. Assume first that $T_+(u_{0,c}, u_{1,c}) = 1$. Assume, to argue by contradiction, that (say) $\int \partial_{x_1}(u_{0,c}) u_{1,c} = \gamma > 0$. Recall that, in this situation, $\operatorname{supp} u_c, \partial_t u_c \subset B(0, 1-t)$, $0 < t < 1$. For convenience, set $u(x,t) = u_c(x, 1+t)$, $-1 < t < 0$, which is supported in $B(0, |t|)$. For $0 < a < 1/4$, we consider the Lorentz transformation

$$z_a(x_1, x', t) = u \left(\frac{x_1 - at}{\sqrt{1-a^2}}, x', \frac{t - ax_1}{\sqrt{1-a^2}} \right),$$

and we fix our attention on $-1/2 \le t < 0$. In that region, the previous lemma and the following comment show, in conjunction with the support property of u, that z_a is a solution in the energy space of (4.3.1). An easy calculation shows that $\operatorname{supp} z_a(\,\cdot\,,t) \subset B(0,|t|)$, so that 0 is the final time of existence for z_a. A lengthy calculation shows that

$$\lim_{a\downarrow 0} \frac{E((z_a(\,\cdot\,,-1/2),\partial_t z_a(\,\cdot\,,-1/2))) - E((u_{0,c},u_{1,c}))}{a} = -\gamma$$

and that, for some $t_0 \in [-1/2,-1/4]$, $\int |\nabla z_a(t_0)|^2 < \int |\nabla W|^2$, for a small (by integration in t_0 and a change of variables, together with the variational estimates for u_c). But, since $E((u_{0,c},u_{1,c})) = E_c$, for a small this contradicts the definition of E_c, since the final time of existence of z_a is finite.

In the case when $T_+(u_{0,c},u_{1,c}) = +\infty$, $\lambda(t) \ge A_0 > 0$, the finiteness of the energy of z_a is unclear, because of the lack of the support property. We instead do a renormalization. We first rescale u_c and consider, for R large, $u_R(x,t) = R^{N-2/2} u_c(Rx, Rt)$, and for a small,

$$z_{a,R}(x_1,x',t) = u_R\left(\frac{x_1 - at}{\sqrt{1-a^2}},\; x',\; \frac{t - ax_1}{\sqrt{1-a^2}} \right).$$

We assume, as before, that $\int \partial_{x_1}(u_{0,c})u_{1,c} = \gamma > 0$ and hope to obtain a contradiction. We prove, by integration in $t_0 \in (1,2)$, that if $h(t_0) = \theta(x)z_{a,R}(x_1,x',t_0)$, with θ a fixed cut-off function, for some a_1 small and R large, we have, for some $t_0 \in (1,2)$, that

$$E((h(t_0),\partial_t h(t_0))) < E_c - \frac{1}{2}\gamma a_1$$

and

$$\int |\nabla h(t_0)|^2 < \int |\nabla W|^2.$$

We then let v be the solution of (4.3.1) with data $h(\,\cdot\,,t_0)$. By the properties of E_c, we know that $\|v\|_{S(-\infty,+\infty)} \le g(\frac{1}{2}\gamma a_1)$, for R large. But, since $\|u_c\|_{S(0,+\infty)} = +\infty$, we have that

$$\|u_R\|_{L^{2(N+1)/N-2}_{[0,1]} L^{2(N+1)/N-2}_{\{|x|<1\}}} \xrightarrow[R\to\infty]{} \infty.$$

But, by finite speed of propagation, we have that $v = z_{a,R}$ on a large set and, after a change of variables to undo the Lorentz transformation, we reach a contradiction from these two facts. $\square$

From all this we see that, to prove Theorem 2, it suffices to show:

Step 3: Rigidity Theorem.

Theorem 8 (Rigidity). *Assume that $E((u_0,u_1)) < E((W,0))$, $\int |\nabla u_0|^2 < \int |\nabla W|^2$. Let u be the corresponding solution of (4.3.1), and let $I_+ = [0, T_+((u_0,u_1)))$. Suppose that:*

a) $\int \nabla u_0 u_1 = 0.$

b) *There exist* $x(t)$, $\lambda(t)$, $t \in [0, T_+((u_0, u_1)))$ *such that*

$$K = \left\{ v(x,t) = \left(\frac{1}{\lambda(t)^{N-2/2}} u_c \left(\frac{x - x(t)}{\lambda(t)}, t \right), \frac{1}{\lambda(t)^{N/2}} \partial_t u_c \left(\frac{x - x(t)}{\lambda(t)}, t \right) \right) : t \in I_+ \right\}$$

has compact closure in $\dot{H}^1 \times L^2$.

c) $x(t)$, $\lambda(t)$ *are continuous,* $\lambda(t) > 0$. *If* $T_+(u_0, u_1) < \infty$, *we have* $\lambda(t) > C/(T_+ - t)$, $\operatorname{supp} u, \partial_t u \subset B(0, T_+ - t)$, *and if* $T_+(u_0, u_1) = +\infty$, *we have* $x(0) = 0$, $\lambda(0) = 1$, $\lambda(t) \geq A_0 > 0$.

Then $T_+(u_0, u_1) = +\infty$, $u \equiv 0$.

Clearly this Rigidity Theorem provides the contradiction that concludes the proof of Theorem 2.

Proof of the Rigidity Theorem. For the proof we need some known identities (see [24], [42]).

Lemma 6. *Let*

$$r(R) = r(t, R) = \int_{|x| \geq R} \left\{ |\nabla u|^2 + |\partial_t u|^2 + |u|^{2^*} + \frac{|u|^2}{|x|^2} \right\} dx.$$

Let u *be a solution of* (4.3.1), $t \in I$, $\phi_R(x) = \phi(x/R)$, $\psi_R(x) = x\phi(x/R)$, *where* ϕ *is in* $C_0^\infty(B_2)$, $\phi \equiv 1$ *on* $|x| \leq 1$. *Then:*

i) $\partial_t \left(\int \psi_R \nabla u \, \partial_t u \right) = -\frac{N}{2} \int (\partial_t u)^2 + \frac{N-2}{2} \int \left[|\nabla u|^2 - |u|^{2^*} \right] + \mathcal{O}(r(R)).$

ii) $\partial_t \left(\int \phi_R \nabla u \, \partial_t u \right) = \int (\partial_t u)^2 - \int |\nabla u|^2 + \int |u|^{2^*} + \mathcal{O}(r(R)).$

iii) $\partial_t \left(\int \psi_R \left\{ \frac{1}{2} |\nabla u|^2 + \frac{1}{2} (\partial_t u)^2 - \frac{1}{2^*} |u|^{2^*} \right\} \right) = -\int \nabla u \, \partial_t u + \mathcal{O}(r(R)).$

We start out the proof of case 1, $T_+((u_0, u_1)) = +\infty$, by observing that, if $(u_0, u_1) \neq (0, 0)$ and $E = E((u_0, u_1))$, then, from our variational estimates, $E > 0$ and

$$\sup_{t>0} \|(\nabla u(t), \partial_t u(t))\|_{\dot{H}^1 \times L^2} \leq CE.$$

We also have

$$\int |\nabla u(t)|^2 - |u(t)|^{2^*} \geq C \int |\nabla u(t)|^2, \quad t > 0$$

and

$$\frac{1}{2} \int (\partial_t u(t))^2 + \frac{1}{2} \int \left[|\nabla u(t)|^2 - |u(t)|^{2^*} \right] \geq CE, \quad t > 0.$$

The compactness of $\overline{K}$ and the fact that $\lambda(t) \geq A_0 > 0$ show that, given $\epsilon > 0$, we can find $R_0(\epsilon) > 0$ so that, for all $t > 0$, we have

$$\int_{\left|x+\frac{x(t)}{\lambda(t)}\right| \geq R(\epsilon)} |\partial_t u|^2 + |\nabla u|^2 + \frac{|u|^2}{|x|^2} + |u|^{2^*} \leq \epsilon E.$$

The proof of this case is accomplished through two lemmas.

Lemma 7. *There exist $\epsilon_1 > 0$, $C > 0$ such that, if $0 < \epsilon < \epsilon_1$, if $R > 2R_0(\epsilon)$, there exists $t_0 = t_0(R, \epsilon)$ with $0 < t_0 \leq CR$, such that for $0 < t < t_0$, we have $\left|\frac{x(t)}{\lambda(t)}\right| < R - R_0(\epsilon)$ and $\left|\frac{x(t)}{\lambda(t)}\right| = R - R_0(\epsilon)$.*

Note that in the radial case, since we can take $x(t) \equiv 0$, a contradiction follows directly from Lemma 7. This will be the analog of the local virial identity proof for the corresponding case of (4.2.1). For the non-radial case we also need:

Lemma 8. *There exist $\epsilon_2 > 0$, $R_1(\epsilon) > 0$, $C_0 > 0$, so that if $R > R_1(\epsilon)$, for $0 < \epsilon < \epsilon_2$, we have $t_0(R, \epsilon) \geq C_0 R/\epsilon$, where t_0 is as in Lemma 7.*

From Lemma 7 and Lemma 8 we have, for $0 < \epsilon < \epsilon_1$, $R > 2R_0(\epsilon)$, $t_0(R, \epsilon) \leq CR$, while for $0 < \epsilon < \epsilon_2$, $R > R_1(\epsilon)$, $t_0(R, \epsilon) \geq C_0 R/\epsilon$. This is clearly a contradiction for ϵ small.

Proof of Lemma 7. Since $x(0) = 0$, $\lambda(0) = 1$; if not, we have for all $0 < t < CR$, with C large, that $\left|\frac{x(t)}{\lambda(t)}\right| < R - R_0(\epsilon)$. Let

$$z_R(t) = \int \psi_R \nabla u\, \partial_t u + \left(\frac{N}{2} - \frac{1}{2}\right) \int \phi_R u\, \partial_t u.$$

Then

$$z_R'(t) = -\frac{1}{2} \int (\partial_t u)^2 - \frac{1}{2} \int \left[|\nabla u|^2 - |u|^{2^*}\right] + \mathcal{O}(r(R)).$$

But, for $|x| > R$, $0 < t < CR$, we have $\left|x + \frac{x(t)}{\lambda(t)}\right| \geq R_0(\epsilon)$ so that $|r(R)| \leq \tilde{C}\epsilon E$. Thus, for ϵ small, $z_R'(t) \leq -\tilde{C}E/2$. By our variational estimates, we also have $|z_R(T)| \leq C_1 R E$. Integrating in t we obtain $CR\tilde{C}E/2 \leq 2C_1 R E$, which is a contradiction for C large. $\qquad\square$

Proof of Lemma 8. For $0 \leq t \leq t_0$, set

$$y_R(t) = \int \psi_R \left\{\frac{1}{2}(\partial_t u)^2 + \frac{1}{2}|\nabla u|^2 - \frac{1}{2^*}|u|^{2^*}\right\}.$$

For $|x| > R$, $\left|x + \frac{x(t)}{\lambda(t)}\right| \geq R_0(\epsilon)$, so that, since $\int \nabla u_0 u_1 = 0 = \int \nabla u(t)\, \partial_t u(t)$, $y'(R) = \mathcal{O}(r(R))$, and hence $|y_R(t_0) - y_R(0)| \leq \tilde{C}\epsilon E t_0$. However,

$$|y_R(0)| \leq \tilde{C}R_0(\epsilon)E + \mathcal{O}(Rr(R_0(\epsilon))) \leq \tilde{C}E[R_0(\epsilon) + \epsilon R].$$

Also,

$$|y_R(t_0)| \geq \left| \int_{\left|x + \frac{x(t_0)}{\lambda(t_0)}\right| \leq R_0(\epsilon)} \psi_R \left\{ \frac{1}{2}(\partial_t u)^2 + \frac{1}{2}|\nabla u|^2 - \frac{1}{2^*}|u|^{2^*} \right\} \right|$$
$$- \left| \int_{\left|x + \frac{x(t_0)}{\lambda(t_0)}\right| > R_0(\epsilon)} \psi_R \left\{ \frac{1}{2}(\partial_t u)^2 + \frac{1}{2}|\nabla u|^2 - \frac{1}{2^*}|u|^{2^*} \right\} \right|.$$

In the first integral, $|x| \leq R$, so that $\psi_R(x) = x$. The second integral is bounded by $MR\epsilon E$. Thus,

$$|y_R(t_0)| \geq \left| \int_{\left|x + \frac{x(t_0)}{\lambda(t_0)}\right| \leq R_0(\epsilon)} x \left\{ \frac{1}{2}(\partial_t u)^2 + \frac{1}{2}|\nabla u|^2 - \frac{1}{2^*}|u|^{2^*} \right\} \right| - MR\epsilon E.$$

The integral on the right equals

$$- \frac{x(t_0)}{\lambda(t_0)} \int_{\left|x + \frac{x(t_0)}{\lambda(t_0)}\right| \leq R_0(\epsilon)} \left\{ \frac{1}{2}(\partial_t u)^2 + \frac{1}{2}|\nabla u|^2 - \frac{1}{2^*}|u|^{2^*} \right\}$$
$$+ \int_{\left|x + \frac{x(t_0)}{\lambda(t_0)}\right| \leq R_0(\epsilon)} \left(x + \frac{x(t_0)}{\lambda(t_0)} \right) \left\{ \frac{1}{2}(\partial_t u)^2 + \frac{1}{2}|\nabla u|^2 - \frac{1}{2^*}|u|^{2^*} \right\},$$

so that its absolute value is greater than or equal to

$$(R_0 - R_0(\epsilon))E - \tilde{C}(R - R_0(\epsilon))\epsilon E - \tilde{C}R_0(\epsilon)E.$$

Thus,

$$|y_R(t_0)| \geq E(R - R_0(\epsilon))[1 - \tilde{C}\epsilon] - \tilde{C}R_0(\epsilon)E - MR\epsilon E \geq ER/4,$$

for R large, ϵ small. But then $ER/4 - \tilde{C}E[R_0(\epsilon) + \epsilon R] \leq \tilde{C}\epsilon Et_0$, which yields the lemma for ϵ small, R large. $\square$

We next turn to the case 2, $T_+((u_0, u_1)) = 1$, with $\operatorname{supp} u, \partial_t u \subset B(0, 1 - t)$, $\lambda(t) \geq C/1 - t$. For (4.3.1) we cannot use the conservation of the L^2 norm as in the (4.2.1) case and a new approach is needed. The first step is:

Lemma 9. *Let u be as in the Rigidity Theorem, with $T_+((u_0, u_1)) = 1$. Then there exists $C > 0$ so that $\lambda(t) \leq C/1 - t$.*

Proof. If not, we can find $t_n \uparrow 1$ so that $\lambda(t_n)(1 - t_n) \to +\infty$. Let

$$z(t) = \int x \nabla u \, \partial_t u + \left(\frac{N}{2} - \frac{1}{2} \right) \int u \, \partial_t u,$$

where we recall that z is well defined since $\operatorname{supp} u, \partial_t u \subset B(0, 1 - t)$. Then, for $0 < t < 1$, we have

$$z'(t) = -\frac{1}{2} \int (\partial_t u)^2 - \frac{1}{2} \int |\nabla u|^2 - |u|^{2^*}.$$

By our variational estimates, $E((u_0, u_1)) = E > 0$ and

$$\sup_{0 < t < 1} \|(u(t), \partial_t u)\|_{\dot{H}^1 \times L^2} \leq CE$$

and $z'(t) \leq -CE$, for $0 < t < 1$. From the support properties of u, it is easy to see that $\lim_{t \uparrow 1} z(t) = 0$, so that, integrating in t, we obtain

$$z(t) \geq CE(1 - t), \quad 0 \leq t < 1.$$

We will next show that $z(t_n)/(1 - t_n) \xrightarrow[n \to \infty]{} 0$, yielding a contradiction. Because $\int \nabla u(t) \, \partial_t u(t) = 0$, $0 < t < 1$, we have

$$\frac{z(t_n)}{1 - t_n} = \int \frac{(x + x(t_n)/\lambda(t_n)) \nabla u \, \partial_t u}{1 - t_n} + \left(\frac{N}{2} - \frac{1}{2}\right) \int \frac{u \, \partial_t u}{1 - t_n}.$$

Note that, for $\epsilon > 0$ given, we have

$$\int_{|x + \frac{x(t_n)}{\lambda(t_n)}| \leq \epsilon(1 - t_n)} \left| x + \frac{x(t_n)}{\lambda(t_n)} \right| |\nabla u(t_n)| |\partial_t u(t_n)| + |u(t_n)| |\partial_t u(t_n)| \leq C_\epsilon E(1 - t_n).$$

Next we will show that $|x(t_n)/\lambda(t_n)| \leq 2(1 - t_n)$. If not, $B(-x(t_n)/\lambda(t_n), (1 - t_n)) \cap B(0, (1 - t_n)) = \emptyset$, so that

$$\int_{B(-x(t_n)/\lambda(t_n), (1-t_n))} |\nabla u(t_n)|^2 + |\partial_t u(t_n)|^2 = 0,$$

while

$$\int_{|x + \frac{x(t_n)}{\lambda(t_n)}| \geq (1-t_n)} |\nabla u(t_n)|^2 + |\partial_t u(t_n)|^2 = \int_{|y| \geq \lambda(t_n)(1-t_n)} \left| \nabla u \left(\frac{y - x(t_n)}{\lambda(t_n)}, t_n \right) \right|^2$$

$$+ \left| \partial_t u \left(\frac{y - x(t_n)}{\lambda(t_n)}, t_n \right) \right|^2 \frac{dy}{\lambda(t_n)^N} \xrightarrow[n \to \infty]{} 0,$$

which contradicts $E > 0$. Then

$$\frac{1}{1 - t_n} \int_{|x + \frac{x(t_n)}{\lambda(t_n)}| \geq \epsilon(1-t_n)} \left| x + \frac{x(t_n)}{\lambda(t_n)} \right| |\nabla u(t_n)| |\partial_t u(t_n)|$$

$$\leq 3 \int_{|x + \frac{x(t_n)}{\lambda(t_n)}| \geq \epsilon(1-t_n)} |\nabla u(t_n)| |\partial_t u(t_n)|$$

$$= 3 \int_{|y| \geq \epsilon(1-t_n)\lambda(t_n)} \left| \nabla u \left(\frac{y - x(t_n)}{\lambda(t_n)}, t_n \right) \right| \left| \partial_t u \left(\frac{y - x(t_n)}{\lambda(t_n)}, t_n \right) \right| \frac{dy}{\lambda(t_n)^N}$$

$$\xrightarrow[n \to \infty]{} 0$$

because of the compactness of $\overline{K}$ and the fact that $\lambda(t_n)(1 - t_n) \to \infty$. Arguing similarly for $\int \frac{u \, \partial_t u}{1 - t_n}$, using Hardy's inequality (centered at $-x(t_n)/\lambda(t_n)$), the proof is concluded. $\qquad \square$

Proposition 5. *Let u be as in the Rigidity Theorem, with $T_+((u_0, u_1)) = 1$, supp u, $\partial_t u \subset B(0, 1-t)$. Then*

$$K = \left((1-t)^{N-2/2} u((1-t)x, t), (1-t)^{N-2/2} \partial_t u((1-t)x, t) \right)$$

is precompact in $\dot{H}^1(\mathbb{R}^N) \times L^2(\mathbb{R}^N)$.

Proof.

$$\left\{ \vec{v}(x, t) = (1-t)^{\frac{N}{2}} \left(\nabla u((1-t)(x - x(t)), t), \partial_t u((1-t)(x - x(t)), t) \right), 0 \leq t < 1 \right\}$$

has compact closure in $L^2(\mathbb{R}^N)^{N+1}$, since we have $c_0 \leq (1-t)\lambda(t) \leq c_1$ and if $\overline{K}$ is compact in $L^2(\mathbb{R}^N)^{N+1}$,

$$K_1 = \left\{ \lambda^{N/2} \vec{v}(\lambda x) \ : \ \vec{v} \in K, c_0 \leq \lambda \leq c_1 \right\}$$

also has $\overline{K}_1$ compact. Let now

$$\tilde{v}(x, t) = (1-t)^{N/2} \left(\nabla u((1-t)x, t), \partial_t u((1-t)x, t) \right),$$

so that $\tilde{v}(x, t) = \vec{v}(x + x(t), t)$. Since supp $\vec{v}(\,\cdot\,, t) \subset \{x : |x - x(t)| \leq 1\}$ and $E > 0$, the fact that $\{\vec{v}(\,\cdot\,, t)\}$ is compact implies that $|x(t)| \leq M$. But if $K_2 = \{\vec{v}(x + x_0, t) : |x_0| \leq M\}$, then $\overline{K}_2$ is compact, giving the proposition. $\qquad\square$

At this point we introduce a new idea, inspired by the works of Giga–Kohn [12] in the parabolic case and Merle–Zaag [36] in the hyperbolic case, who studied the equations $(\partial_t^2 - \triangle)u - |u|^{p-1}u = 0$, for $1 < p < \frac{4}{N-1} + 1$, in the radial case. In our case, $p = \frac{4}{N-2} + 1 > \frac{4}{N-1} + 1$. We thus introduce self-similar variables. Thus, we set $y = x/1 - t$, $s = \log 1/1 - t$ and define

$$w(y, s; 0) = (1-t)^{N-2/2} u(x, t) = e^{-s(N-2)/2} u(e^{-s}y, 1 - e^{-s}),$$

which is defined for $0 \leq s < \infty$ with supp $w(\,\cdot\,, s; 0) \subset \{|y| \leq 1\}$. We will also consider, for $\delta > 0$, $u_\delta(x, t) = u(x, t + \delta)$ which also solves (4.3.1) and its corresponding w, which we will denote by $w(y, s; \delta)$. Thus, we set $y = x/1 + \delta - t$, $s = \log 1/1 + \delta - t$ and

$$w(y, s; \delta) = (1 + \delta - t)^{N-2/2} u(x, t) = e^{-s(N-2)/2} u(e^{-s}y, 1 + \delta - e^{-s}).$$

Here $w(y, s; \delta)$ is defined for $0 \leq s < -\log \delta$ and we have

$$\text{supp } w(\,\cdot\,, s; \delta) \subset \left\{ |y| \leq \frac{e^{-s} - \delta}{e^{-s}} = \frac{1-t}{1 + \delta - t} \leq 1 - \delta \right\}.$$

The w solve, where they are defined, the equation

$$\partial_s^2 w = \frac{1}{\rho} \text{div} \left(\rho \nabla w - \rho(y \cdot \nabla w)y \right) - \frac{N(N-2)}{4} w$$

$$+ |w|^{4/N-2} w - 2y \cdot \nabla \partial_s w - (N-1)\partial_s w,$$

where $\rho(y) = (1 - |y|^2)^{-1/2}$.

Note that the elliptic part of this operator degenerates. In fact,

$$\frac{1}{\rho}\mathrm{div}\ (\rho\nabla w - \rho(y\cdot\nabla w)y) = \frac{1}{\rho}\mathrm{div}\ (\rho(I - y\otimes y)\nabla w),$$

which is elliptic with smooth coefficients for $|y| < 1$, but degenerates at $|y| = 1$.

Here are some straightforward bounds on $w(\,\cdot\,;\delta)$ ($\delta > 0$): $w \in H_0^1(B_1)$ with

$$\int_{B_1} |\nabla w|^2 + |\partial_s w|^2 + |w|^{2^*} \leq C.$$

Moreover, by Hardy's inequality for $H_0^1(B_1)$ functions [6],

$$\int_{B_1} \frac{|w(y)|^2}{(1 - |y|^2)^2} \leq C.$$

These bounds are uniform in $\delta > 0$, $0 < s < -\log\delta$. Next, following [36], we introduce an energy, which will provide us with a Lyapunov functional for w.

$$\tilde{E}(w(s;\delta)) = \int_{B_1} \frac{1}{2}\left\{(\partial_s w)^2 + |\nabla w|^2 - (y\cdot\nabla w)^2\right\}\frac{dy}{(1 - |y|^2)^{1/2}}$$
$$+ \int_{B_1}\left\{\frac{N(N-2)}{8}w^2 - \frac{N-2}{2N}|w|^{2^*}\right\}\frac{dy}{(1 - |y|^2)^{1/2}}.$$

Note that this is finite for $\delta > 0$. We have:

Lemma 10. *For $\delta > 0$, $0 < s_1 < s_2 < \log 1/\delta$,*

i) $\tilde{E}(w(s_2)) - \tilde{E}(w(s_1)) = \displaystyle\int_{s_1}^{s_2}\int_{B_1}\frac{(\partial_s w)^2}{(1 - |y|^2)^{3/2}}\,ds\,dy$, *so that $\tilde{E}$ is increasing.*

ii) $\displaystyle\frac{1}{2}\int_{B_1}\left[(\partial_s w)\cdot w - \frac{1+N}{2}w^2\right]\frac{dy}{(1 - |y|^2)^{1/2}}\bigg|_{s_1}^{s_2}$

$\displaystyle = -\int_{s_1}^{s_2}\tilde{E}(w(s))ds + \frac{1}{N}\int_{s_1}^{s_2}\int_{B_1}\frac{|w|^{2^*}}{(1 - |y|^2)^{1/2}}\,ds\,dy$

$\displaystyle + \int_{s_1}^{s_2}\int_{B_1}\left\{(\partial_s w)^2 + \partial_s wy\cdot\nabla w + \frac{\partial_s ww|y|^2}{1 - |y|^2}\right\}\frac{dy}{(1 - |y|^2)^{1/2}}.$

iii) $\displaystyle\lim_{s\to\log 1/\delta}\tilde{E}(w(s)) = E((u_0, u_1)) = E$, *so that, by part i), $\tilde{E}(w(s)) \leq E$ for $0 \leq s < \log 1/\delta$.*

The proof is computational; see [24]. Our first improvement over this is:

Lemma 11. $\displaystyle\int_0^1\int_{B_1}\frac{(\partial_s w)^2}{1 - |y|^2}\,dy\,ds \leq C\log 1/\delta.$

Proof. Notice that

$$-2 \int \frac{(\partial_s w)^2}{1 - |y|^2} = \frac{d}{ds} \Bigg\{ \int \bigg[\frac{1}{2} (\partial_s w)^2 + \frac{1}{2} \left(|\nabla w|^2 - (y \cdot \nabla w)^2 \right) \bigg.$$

$$+ \frac{(N-2)N}{8} w^2 - \frac{N-2}{2N} |w|^{2^*} \bigg] \left[-\log(1 - |y|^2) \right] dy$$

$$+ \int \left[\log(1 - |y|^2) + 2 \right] y \cdot \nabla w \, \partial_s w - \log(1 - |y|^2)(\partial_s w)^2$$

$$- 2 \int (\partial_s w)^2 \Bigg\}.$$

We next integrate in s, between 0 and 1, and drop the next to last term by sign. The proof is finished by using Cauchy–Schwartz and the support property of $w(\,\cdot\,;\delta)$. $\qquad \square$

Corollary 4. a) $\displaystyle \int_0^1 \int_{B_1} \frac{|w|^{2^*}}{(1 - |y|^2)^{1/2}} \, dy \, ds \leq C (\log 1/\delta)^{1/2}.$

b) $\tilde{E}(w(1)) \geq -C(\log 1/\delta)^{1/2}.$

Proof. Part a) follows from ii), iii) above, Cauchy–Schwartz and Lemma 11. Note that we obtain the power $1/2$ on the right-hand side by Cauchy–Schwartz. Part b) follows from i) and the fact that

$$\int_0^1 \tilde{E}(w(s)) \, ds \geq -C (\log 1/\delta)^{1/2},$$

which is a consequence of the definition of $\tilde{E}$ and a). $\qquad \square$

Our next improvement is:

Lemma 12. $\displaystyle \int_1^{\log 1/\delta} \int_{B_1} \frac{(\partial_s w)^2}{(1 - |y|^2)^{3/2}} \leq C (\log 1/\delta)^{1/2}.$

Proof. Use i), iii) and the bound b) in Corollary 4. $\qquad \square$

Corollary 5. *There exists* $\bar{s}_\delta \in \left(1, (\log 1/\delta)^{3/4} \right)$ *such that*

$$\int_{\bar{s}_\delta}^{\bar{s}_\delta + (\log 1/\delta)^{1/8}} \int_{B_1} \frac{(\partial_s w)^2}{(1 - |y|^2)^{3/2}} \leq \frac{C}{(\log 1/\delta)^{1/8}}.$$

Proof. Split $\left(1, (\log 1/\delta)^{3/4} \right)$ into disjoint intervals of length $(\log 1/\delta)^{1/8}$. Their number is $(\log 1/\delta)^{5/8}$ and $\frac{5}{8} - \frac{1}{8} = \frac{1}{2}$. $\qquad \square$

Note that, in Corollary 5, the length of the s interval tends to infinity, while the bound goes to zero. It is easy to see that if $\bar{s}_\delta \in \left(1, (\log 1/\delta)^{3/4}\right)$, and $\bar{s}_\delta = -\log(1 + \delta - \bar{t}_\delta)$, then

$$\left| \frac{1 - \bar{t}_\delta}{1 + \delta - \bar{t}_\delta} - 1 \right| \leq C\delta^{1/4},$$

which goes to 0 with δ. From this and the compactness of $\overline{K}$, we can find $\delta_j \to 0$, so that $w(y, \bar{s}_{\delta_j} + s; \delta_j)$ converges, for $s \in [0, S]$ to $w^*(y, s)$ in $C([0, S]; \dot{H}_0^1 \times L^2)$, and w^* solves our self-similar equation in $B_1 \times [0, S]$. Corollary 5 shows that w^* must be independent of s. Also, the fact that $E > 0$ and our coercivity estimates show that $w^* \not\equiv 0$. (See [24] for the details.) Thus, $w^* \in H_0^1(B_1)$ solves the (degenerate) elliptic equation

$$\frac{1}{\rho} \operatorname{div} \left(\rho \nabla w^* - \rho(y \cdot \nabla w^*)y \right) - \frac{N(N-2)}{4} w^* + |w^*|^{4/N-2} w^* = 0,$$

$$\rho(y) = (1 - |y|^2)^{-1/2}.$$

We next point out that w^* satisfies the additional (crucial) estimates:

$$\int_{B_1} \frac{|w^*|^{2^*}}{(1 - |y|^2)^{1/2}} + \int_{B_1} \frac{[|\nabla w^*|^2 - (y \cdot \nabla w^*)^2]}{(1 - |y|^2)^{1/2}} < \infty.$$

Indeed, for the first estimate it suffices to show that, uniformly in j large, we have

$$\int_{\bar{s}_{\delta_j}}^{\bar{s}_{\delta_j} + \delta} \int_{B_1} \frac{|w(y, s; \delta_j)|^{2^*}}{(1 - |y|^2)^{1/2}} \, dy \, ds \leq C,$$

which follows from ii) above, together with the choice of $\bar{s}_{\delta_j}$, by Corollary 5, Cauchy–Schwartz and iii). The proof of the second estimate follows from the first one, iii) and the formula for $\tilde{E}$.

The conclusion of the proof is obtained by showing that a w^* in $H_0^1(B_1)$, solving the degenerate elliptic equation with the additional bounds above, must be zero. This will follow from a unique continuation argument. Recall that, for $|y| \leq 1 - \eta_0$, $\eta_0 > 0$, the linear operator is uniformly elliptic, with smooth coefficients and that the nonlinearity is critical. An argument of Trudinger's [51] shows that w^* is bounded on $\{|y| \leq 1 - \eta_0\}$ for each $\eta_0 > 0$. Thus, if we show that $w^* \equiv 0$ near $|y| = 1$, the standard Carleman unique continuation principle [19] will show that $w^* \equiv 0$.

Near $|y| = 1$, our equation is modeled (in variables $z \in \mathbb{R}^{N-1}$, $r \in \mathbb{R}$, $r > 0$, near $r = 0$) by

$$r^{1/2} \partial_r (r^{1/2} \partial_r w^*) + \triangle_z w^* + c w^* + |w^*|^{4/N-2} w^* = 0.$$

Our information on w^* translates into $w^* \in H_0^1((0, 1] \times (|z| < 1))$ and our crucial additional estimates are:

$$\int_0^1 \int_{|z|<1} |w^*(r, z)|^{2^*} \frac{dr}{r^{1/2}} \, dz + \int_0^1 \int_{|z|<1} |\nabla_z w^*(r, z)|^2 \frac{dr}{r^{1/2}} \, dz < \infty.$$

To conclude, we take advantage of the degeneracy of the equation. We "desingularize" the problem by letting $r = a^2$, setting $v(a, z) = w^*(a^2, z)$, so that $\partial_a v(a, z) = 2r^{1/2} \partial_r w^*(r, z)$. Our equation becomes:

$$\partial_a^2 v + \triangle_z v + cv + |v|^{4/N-2} v = 0, \quad 0 < a < 1, \quad |z| < 1, \quad v|_{a=0} = 0,$$

and our bounds give:

$$\int_0^1 \int_{|z|<1} |\nabla_z v(a, z)|^? \, da \, dz = \int_0^1 \int_{|z|<1} |\nabla_z w^*(r, z)|^? \, \frac{dr}{r^{1/2}} \, dz < \infty,$$

$$\int_0^1 \int_{|z|<1} |\partial_a v(a, z)|^2 \, \frac{da}{a} \, dz = \int_0^1 \int_{|z|<1} |\partial_r w^*(r, z)|^2 \, dr \, dz < \infty.$$

Thus, $v \in H_0^1((0, 1] \times B_1)$, but in addition $\partial_a v(a, z)|_{a=0} \equiv 0$. We then extend v by 0 to $a < 0$ and see that the extension is an H^1 solution to the same equation. By Trudinger's argument, it is bounded. But since it vanishes for $a < 0$, by Carleman's unique continuation theorem, $v \equiv 0$. Hence, $w^* \equiv 0$, giving our contradiction. $\qquad \square$

Bibliography

[1] T. Aubin. Équations différentielles non linéaires et problème de Yamabe concernant la courbure scalaire. *J. Math. Pures Appl. (9)*, 55(3):269–296, 1976.

[2] H. Bahouri and P. Gérard. High frequency approximation of solutions to critical nonlinear wave equations. *Amer. J. Math.*, 121(1):131–175, 1999.

[3] H. Bahouri and J. Shatah. Decay estimates for the critical semilinear wave equation. *Ann. Inst. H. Poincaré Anal. Non Linéaire*, 15(6):783–789, 1998.

[4] J. Bourgain. Global wellposedness of defocusing critical nonlinear Schrödinger equation in the radial case. *J. Amer. Math. Soc.*, 12(1):145–171, 1999.

[5] H. Brézis and J.-M. Coron. Convergence of solutions of H-systems or how to blow bubbles. *Arch. Rational Mech. Anal.*, 89(1):21–56, 1985.

[6] H. Brézis and M. Marcus. Hardy's inequalities revisited. *Ann. Scuola Norm. Sup. Pisa Cl. Sci. (4)*, 25(1-2):217–237 (1998), 1997. Dedicated to Ennio De Giorgi.

[7] T. Cazenave and F. B. Weissler. The Cauchy problem for the critical nonlinear Schrödinger equation in H^s. *Nonlinear Anal.*, 14(10):807–836, 1990.

[8] J. Colliander, M. Keel, G. Staffilani, H. Takaoka, and T. Tao. Global well-posedness and scattering for the energy-critical nonlinear Schrödinger equation in $\mathbb{R}^3$. *Ann. of Math. (2)*, 167(3):767–865, 2008.

[9] R. Côte, C. Kenig, and F. Merle. Scattering below critical energy for the radial 4D Yang–Mills equation and for the 2D corotational wave map system. *Commun. Math. Phys.*, 284(1):203–225, 2008.

[10] T. Duyckaerts, J. Holmer, and S. Roudenko. Scattering for the non-radial 3D cubic nonlinear Schrödinger equation. *Math. Res. Lett.*, 15:1233–1250, 2008.

[11] L. Escauriaza, G. A. Serëgin, and V. Sverak. $L_{3,\infty}$-solutions of Navier–Stokes equations and backward uniqueness. *Russ. Math. Surv.*, 58(2):211–250, 2003.

[12] Y. Giga and R. V. Kohn. Nondegeneracy of blowup for semilinear heat equations. *Comm. Pure Appl. Math.*, 42(6):845–884, 1989.

[13] J. Ginibre, A. Soffer, and G. Velo. The global Cauchy problem for the critical nonlinear wave equation. *J. Funct. Anal.*, 110(1):96–130, 1992.

[14] J. Ginibre and G. Velo. Generalized Strichartz inequalities for the wave equation. *J. Funct. Anal.*, 133(1):50–68, 1995.

[15] R. T. Glassey. On the blowing up of solutions to the Cauchy problem for nonlinear Schrödinger equations. *J. Math. Phys.*, 18(9):1794–1797, 1977.

[16] M. G. Grillakis. Regularity and asymptotic behaviour of the wave equation with a critical nonlinearity. *Ann. of Math. (2)*, 132(3):485–509, 1990.

[17] M. G. Grillakis. Regularity for the wave equation with a critical nonlinearity. *Comm. Pure Appl. Math.*, 45(6):749–774, 1992.

[18] M. G. Grillakis. On nonlinear Schrödinger equations. *Comm. Partial Differential Equations*, 25(9-10):1827–1844, 2000.

[19] L. Hörmander. *The analysis of linear partial differential operators. III*, volume 274 of *Grundlehren der Mathematischen Wissenschaften [Fundamental Principles of Mathematical Sciences]*. Springer-Verlag, Berlin, 1985. Pseudodifferential operators.

[20] L. Kapitanski. Global and unique weak solutions of nonlinear wave equations. *Math. Res. Lett.*, 1(2):211–223, 1994.

[21] M. Keel and T. Tao. Endpoint Strichartz estimates. *Amer. J. Math.*, 120(5):955–980, 1998.

[22] C. Kenig. Global well-posedness and scattering for the energy critical focusing non-linear Schrödinger and wave equations. Lecture Notes for a mini-course given at "Analyse des équations aux derivées partialles", Evian-les-bains, June 2007.

[23] C. Kenig and F. Merle. Global well-posedness, scattering and blow-up for the energy-critical, focusing, non-linear Schrödinger equation in the radial case. *Invent. Math.*, 166(3):645–675, 2006.

[24] C. Kenig and F. Merle. Global well-posedness, scattering and blow-up for the energy critical focusing non-linear wave equation. *Acta Math.*, 201(2):147–212, 2008.

[25] C. Kenig and F. Merle. Scattering for $\dot{H}^{1/2}$ bounded solutions to the cubic defocusing NLS in 3 dimensions. *Trans. Amer. Math. Soc.*, 362(4):1937–1962, 2010.

[26] C. Kenig, G. Ponce, and L. Vega. Well-posedness and scattering results for the generalized Korteweg-de Vries equation via the contraction principle. *Comm. Pure Appl. Math.*, 46(4):527–620, 1993.

[27] S. Keraani. On the defect of compactness for the Strichartz estimates of the Schrödinger equations. *J. Differential Equations*, 175(2):353–392, 2001.

[28] R. Killip, T. Tao, and M. Vişan. The cubic nonlinear Schrödinger equation in two dimensions with radial data. *J. Eur. Math. Soc.*, 11:1203–1258, 2009.

[29] R. Killip and M. Vişan. The focusing energy-critical nonlinear Schrödinger equation in dimensions five and higher. *Amer. J. Math.*, 132:361–424, 2010.

[30] R. Killip, M. Vişan, and X. Zhang. The mass-critical nonlinear Schrödinger equation with radial data in dimensions three and higher. *Analysis and PDE*, 1:229–266, 2008.

[31] J. Krieger, W. Schlag, and D. Tǎtaru. Renormalization and blow up for charge and equivariant critical wave maps. *Invent. Math.*, 171(3):543–615, 2008.

[32] J. Krieger, W. Schlag, and D. Tǎtaru. Slow blow-up solutions for the $H^1(\mathbb{R}^3)$ critical focusing semi-linear wave equation in $\mathbb{R}^3$. *Duke Math. J.*, 147(1):1–53, 2009.

[33] H. Levine. Instability and nonexistence of global solutions to nonlinear wave equations of the form $Pu_{tt} = -Au + \mathcal{F}(u)$. *Trans. Amer. Math. Soc.*, 192:1–21, 1974.

[34] H. Lindblad and C. Sogge. On existence and scattering with minimal regularity for semilinear wave equations. *J. Funct. Anal.*, 130(2):357–426, 1995.

[35] F. Merle and L. Vega. Compactness at blow-up time for L^2 solutions of the critical nonlinear Schrödinger equation in 2D. *Internat. Math. Res. Notices*, (8):399–425, 1998.

[36] F. Merle and H. Zaag. Determination of the blow-up rate for the semilinear wave equation. *Amer. J. Math.*, 125(5):1147–1164, 2003.

[37] H. Pecher. Nonlinear small data scattering for the wave and Klein–Gordon equation. *Math. Z.*, 185(2):261–270, 1984.

[38] P. Raphaël. Existence and stability of a solution blowing up on a sphere for an L^2-supercritical nonlinear Schrödinger equation. *Duke Math. J.*, 134(2):199–258, 2006.

[39] I. Rodnianski and J. Sterbenz. On the formation of singularities in the critical $O(3)$ σ-model. *Ann. of Math. (2)*, 172(1):187–242, 2010.

[40] E. Ryckman and M. Vişan. Global well-posedness and scattering for the defocusing energy-critical nonlinear Schrödinger equation in $\mathbb{R}^{1+4}$. *Amer. J. Math.*, 129(1):1–60, 2007.

[41] J. Shatah and M. Struwe. Well-posedness in the energy space for semilinear wave equations with critical growth. *Internat. Math. Res. Notices*, (7):303ff., approx. 7 pp. (electronic), 1994.

[42] J. Shatah and M. Struwe. *Geometric wave equations*, volume 2 of *Courant Lecture Notes in Mathematics*. New York University Courant Institute of Mathematical Sciences, New York, 1998.

[43] R. Strichartz. Restrictions of Fourier transforms to quadratic surfaces and decay of solutions of wave equations. *Duke Math. J.*, 44(3):705–714, 1977.

[44] M. Struwe. Globally regular solutions to the u^5 Klein–Gordon equation. *Ann. Scuola Norm. Sup. Pisa Cl. Sci. (4)*, 15(3):495–513 (1989), 1988.

[45] M. Struwe. Equivariant wave maps in two space dimensions. *Comm. Pure Appl. Math.*, 56(7):815–823, 2003. Dedicated to the memory of Jürgen Moser.

[46] G. Talenti. Best constant in Sobolev inequality. *Ann. Mat. Pura Appl. (4)*, 110:353–372, 1976.

[47] T. Tao. Global regularity of wave maps. II. Small energy in two dimensions. *Commun. Math. Phys.*, 224(2):443–544, 2001.

[48] T. Tao. Global well-posedness and scattering for the higher-dimensional energy-critical nonlinear Schrödinger equation for radial data. *New York J. Math.*, 11:57–80 (electronic), 2005.

[49] T. Tao and M. Vişan. Stability of energy-critical nonlinear Schrödinger equations in high dimensions. *Electron. J. Differential Equations*, 118:1–28 (electronic), 2005.

[50] T. Tao, M. Vişan, and X. Zhang. Global well-posedness and scattering for the defocusing mass-critical nonlinear Schrödinger equation for radial data in high dimensions. *Duke Math. J.*, 140(1):165–202, 2007.

[51] N. Trudinger. Remarks concerning the conformal deformation of Riemannian structures on compact manifolds. *Ann. Scuola Norm. Sup. Pisa (3)*, 22:265–274, 1968.

[52] D. Tătaru. On global existence and scattering for the wave maps equation. *Amer. J. Math.*, 123(1):37–77, 2001.

[53] D. Tătaru. Rough solutions for the wave maps equation. *Amer. J. Math.*, 127(2):293–377, 2005.

[54] M. Vişan. The defocusing energy-critical nonlinear Schrödinger equation in higher dimensions. *Duke Math. J.*, 138(2):281–374, 2007.

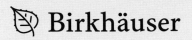

birkhauser-science.com

Advanced Courses in Mathematics – CRM Barcelona (ACM)

Edited by
Carles Casacuberta, Universitat de Barcelona, Spain

Since 1995 the Centre de Recerca Matemàtica (CRM) has organised a number of Advanced Courses at the post-doctoral or advanced graduate level on forefront research topics in Barcelona. The books in this series contain revised and expanded versions of the material presented by the authors in their lectures.

■ **Moerdijk, I. / Toën, B.**, Simplicial Methods for Operads and Algebraic Geometry (2010).
ISBN 8-3-0348-0051-8

This book is an introduction to two higher-categorical topics in algebraic topology and algebraic geometry relying on simplicial methods.

Moerdijk's lectures offer a detailed introduction to dendroidal sets, which were introduced by himself and Weiss as a foundation for the homotopy theory of operads. The theory of dendroidal sets is based on trees instead of linear orders and has many features analogous to the theory of simplicial sets, but it also reveals new phenomena. For example, dendroidal sets admit a closed symmetric monoidal structure related to the Boardman–Vogt tensor product of operads. The lecture notes start with the combinatorics of trees and culminate with a suitable model structure on the category of dendroidal sets. Important concepts are illustrated with pictures and examples.

The lecture series by Toën presents derived algebraic geometry. While classical algebraic geometry studies functors from the category of commutative rings to the category of sets, derived algebraic geometry is concerned with functors from simplicial commutative rings (to allow derived tensor products) to simplicial sets (to allow derived quotients). The central objects are derived (higher) stacks, which are functors satisfying a certain up-to-homotopy descent condition. These lectures provide a concise and focused introduction to this vast subject, glossing over many of the technicalities that make the subject's research literature so overwhelming.

Both sets of lectures assume a working knowledge of model categories in the sense of Quillen. For Toën's lectures, some background in algebraic geometry is also necessary.

■ **Ritoré, M. / Sinestrari, C.**, Mean Curvature Flow and Isoperimetric Inequalities (2010).
ISBN 978-3-0346-0212-9

Geometric flows have many applications in physics and geometry. The mean curvature flow occurs in the description of the interface evolution in certain physical models. This is related to the property that such a flow is the gradient flow of the area functional and therefore appears naturally in problems where a surface energy is minimized. The mean curvature flow also has many geometric applications, in analogy with the Ricci flow of metrics on abstract riemannian manifolds. One can use this flow as a tool to obtain classification results for surfaces satisfying certain curvature conditions, as well as to construct minimal surfaces. Geometric flows, obtained from solutions of geometric parabolic equations, can be considered as an alternative tool to prove isoperimetric inequalities. On the other hand, isoperimetric inequalities can help in treating several aspects of convergence of these flows. Isoperimetric inequalities have many applications in other fields of geometry, like hyperbolic manifolds.

■ **Geroldinger, A. / Ruzsa, I. Z.**, Combinatorial Number Theory and Additive Group Theory (2009).
ISBN 978-3-7643-8961-1

■ **Bertoluzza, S. / Falletta, S. / Russo, G. / Shu, C.-W.**, Numerical Solutions of Partial Differential Equations (2009).
ISBN 978-3-7643-8939-0

■ **Myasnikov, A. / Shpilrain, V. / Ushakov, A.**, Group-based Cryptography (2008).
ISBN 978-3-7643-8826-3

Printing: Ten Brink, Meppel, The Netherlands
Binding: Stürtz, Würzburg, Germany